HOW TO READ
BRIDGES

HOW TO READ
BRIDGES

A crash course spanning the centuries

HERBERT PRESS
LONDON

Edward Denison
& Ian Stewart

First published in Great Britain in 2012
by HERBERT PRESS
An imprint of Bloomsbury Publishing Plc
50 Bedford Square
London WC1B 3DP, UK
www.acblack.com

ISBN: 978-1-4081-7176-9

A CIP catalogue record for this book
is available from the British Library

Colour origination by Ivy Press Reprographics
Printed in China

This book was conceived, designed
and produced by
Ivy Press
210 High Street
Lewes, East Sussex
BN7 2NS, UK
www.ivy-group.co.uk

CREATIVE DIRECTOR Peter Bridgewater
PUBLISHER Jason Hook
EDITORIAL DIRECTOR Caroline Earle
ART DIRECTOR Michael Whitehead
DESIGN JC Lanaway
ILLUSTRATOR Adam Hook
PROJECT EDITOR Jamie Pumfrey
PICTURE MANAGER Katie Greenwood

CONTENTS

INTROD

'Necessity,' so the saying goes, 'is the mother of invention.' For bridge building, this truism is the ultimate raison d'être. From the earliest times when humans needed simply to cross a stream to reach new pastures, to the present day when global trade and communication depend on myriad bridges and complex bridge systems crossing rivers and seas, the need for new bridges has been a constant source of innovation.

Bridges have played a crucial role in humankind's expansion around the globe. From humble origins when bridges were built to fulfil the primitive needs of our ancestors, bridges have helped to define the evolution of settlements, towns, cities and even nations. The outcome of territorial disputes has often hinged on the control of bridges. Throughout history the development of commerce has relied on dependable river crossings. In modern times the expansion of cities and sometimes even the relationship between countries owe a debt to bridges. It is impossible to imagine the world's largest cities such as New York or London without their river crossings, just as ancient cities like Venice or Rome could not have prospered without their bridges.

In recent decades bridge design and construction has witnessed a prodigious growth. The records for the world's longest, tallest, highest, busiest, and heaviest bridges have been repeatedly broken as our knowledge of engineering, materials technologies and construction skills has improved and are constantly satisfying our incessant demand for better means of communication. Today, the latest innovations in design and engineering are employed to construct new bridges big enough to link nations across seas and small enough to create new links and exciting routes in historic cities.

For the engineer, bridges are one of the most stimulating and explicit manifestations of their trade. For the rest of us, most bridges are taken for granted. While some bridges might inspire awe because of their size or form and others might evoke sentimentality because of their age, most bridges are seen with little more reverence than being merely useful or utilitarian. However, being able to read a bridge brings all these structures to life, from the most celebrated examples, to the apparently innocuous.

In this book, reading bridges is arranged into two parts. The first part begins with the basics, examining the materials that have been used to build bridges throughout history, reviewing the many different types of bridges, understanding the variety of uses, and introducing some of the most renowned engineers whose names have become synonymous with bridge design. The second part explores a wide selection of bridges from across the globe through a series of case studies arranged according to bridge type. Each case study helps the reader understand the structure and interpret its distinguishing characteristics.

Understanding why a structure looks the way it does, why it is built the way it is, and why it uses the materials it does, excites the inquisitive mind and opens the untrained eye to a unique field of human ingenuity that can be found all around us. From the busiest city centre to the remotest mountain pass, every bridge has its own story – as long as the viewer is able to read it.

Looking for Clues

Structural form
Although cladding in stone was originally planned for the towers of the George Washington Bridge, they were left unclad, because the exposed structural form was considered more attractive and more economical.

In the modern built environment, bridges are almost unique in presenting an exposed structural form. Old stone or masonry buildings with their buttresses, arches and piers offered a visual honesty that enabled interested observers to appreciate the structure and develop an intuitive understanding of how they stood up. However, as modern building materials have developed, steel, concrete and timber structures are commonly enshrouded in cladding systems that create the aesthetic character of a building while concealing its primary structure.

Bridges similarly have developed rapidly with the material advances of the last two centuries but generally have no need for cladding systems. These structures therefore remain one of the truest combinations of engineering and architecture, presenting a great opportunity to observe and understand structural form.

Direction of bracing

Simple geometric changes can significantly alter the way a structure transfers load back to its supports. In beam truss configurations, if the bracing slopes down towards the centre of the span it carries load in tension. If they slope upwards then they work in compression and must therefore be bigger in cross section to withstand buckling forces. However, cantilever configurations can reverse these rules, as shown above.

Joints

Look out for joints. These can indicate simply supported beams as opposed to continuous beams, and cantilever bridges as opposed to beam or arch bridges. They can also reveal how the bridge was constructed, often with sections fabricated off-site.

Old deck
cross section

New deck
cross section

Hidden foundations

Closer inspection of different arch types can give an indication of how the foundations are working. Deck-arch bridges require foundations that provide resistance to both vertical and lateral components of thrust forces. A tied-arch bridge uses the deck to develop tensile forces meaning that little or no lateral force reaches the foundations. By observing and understanding the superstructure, we can often tell a great deal about foundations that we never get to see.

Authenticity

Maintenance is an important consideration for all bridges. Components have to be replaced or repaired over the working life of the structure. Similarly, many old bridges have had to adapt to changes in usage and increased demands on their capabilities. Some have had their decks modernised, while others are rebuilt to maintain the knowledge of their construction. Look out for non-original elements that give clues about their evolution.

Future Bridges

Although many of the fundamental concepts of bridge design remain the same as they have done for thousands of years, the past two centuries have witnessed a revolution in bridge building. New materials and technologies, and the accumulative experience and improving expertise of designers and engineers have helped to create longer, more efficient and safer structures than ever before.

Bridge building has enjoyed a period of growth that is set to continue into the 21st century. In recent years, world records for the longest, tallest and highest bridges across all categories of type or use have been repeatedly broken. The escalation of record-breaking bridge building is caused in part by the refinement of expertise and the technological tools available to designers and engineers, and in part by the needs of developing nations as they build modern infrastructure. Globalisation and the political and economic union of nations and regions will also continue to fuel bridge building.

The future of bridge building is one of unparalleled potential, as designers and engineers continue to push the boundaries of established knowledge. In the short term the refinement of conventional techniques will create longer and more efficient structures, while in the longer term new materials and technologies will inevitably generate entirely new structural types that will in turn revolutionise bridge design, just as steel and concrete have done in our time.

Third Forth Bridge

This new bridge is designed as a composite cable-stayed and continuous beam structure. The road deck will be carried across the estuary by a combination of three A-frame cable-stayed pylons and concrete piers. The cable-stayed section will create two elevated main navigation channels and two side spans, and the piers will support the approach roads that gently incline from the shoreline to the cable-stayed section. The A-frame pylons will support two sets of cables on each side arranged in a fan-like pattern. These cables will be connected to the sides of the road deck.

Future Bridges

Messina Strait Bridge

The idea of bridging the Messina Strait separating mainland Italy and Sicily goes back as far as the Romans and has not dwindled since. Serious but ultimately fruitless studies were conducted in the late 19th and early 20th centuries, and then renewed interest in the idea in the 1950s led to a competition being held in the 1960s, which also came to nothing. The project finally received state approval in the 1990s, and in 2006 a plan was drafted for the longest and tallest bridge structure ever built. The project faced cancellation but was reinstated in 2009. Although its immediate future remains in doubt, the bridging of the Messina Strait is an idea that will not go away and when finally realised it will represent a milestone in the history of bridge building.

Historic future bridge

The proposed Messina Strait suspension bridge has a single span of 3,300 m (10,800 ft), 60% longer than the current record holder, the Akashi Kaikyo Bridge (see pages 218–219). To provide a sufficient gradient for the suspension cables and a 65 m (213 ft) clearance for shipping, the pylons have to be almost 400 m (1,312 ft) high, exceeding the record holder, the Millau Viaduct (see pages 230–231), by more than 40 m (130 ft). Two pairs of 1.25 m (5 ft) diameter cable are required to carry the 60 m (197 ft) wide deck, comprising six lanes of traffic, two railway lines, two pedestrian lanes and an emergency lane.

The Third Forth Bridge

The Firth of Forth in Scotland is accustomed to record-breaking bridge design. The first bridge across the Forth was the rail bridge (see pages 28 and 181). Opened in 1890, it was the world's longest ever cantilever structure and the first large bridge entirely of steel. The second bridge was the longest suspension bridge outside the United Sates when it opened to vehicular traffic in 1964 (see pages 35 and 56). However, the deterioration of its main cables over half a century has weakened the structure and a new bridge has been designed to replace it. Construction is due to be completed in 2016.

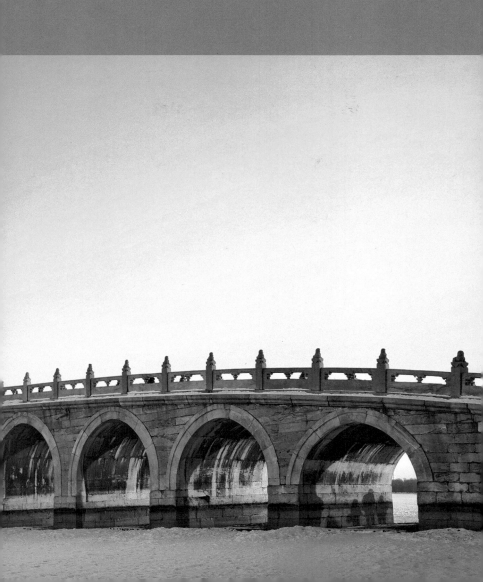

MATERIALS Introduction

Throughout history and across the globe, bridge building has relied on an amazing array of materials. Some, such as wood and stone, have endured for millennia, while others, such as bamboo, roots or vines, are unique to their place. Other materials require specialist manufacturing. Kiln-fired bricks, dating back more than four thousand years, were among the first examples of a manufactured material that was later used for bridge construction.

Over the last two centuries, modern materials and manufacturing processes have revolutionised bridge building. Iron, steel, concrete and even toughened glass have transformed bridge design and enhanced our capacity to bridge longer distances or design more innovative, efficient and safer structures.

It is not just the primary materials from which bridges are built that have contributed to their evolution, but also the secondary materials. Lime mortar was used to bond stones, while iron nails strengthened wooden bridges and iron dovetail joints tied stones together. Most modern bridges rely on the combined and complementary attributes of concrete and steel. Whether in reinforced concrete, prestressed concrete or post-tensioned concrete, steel's tensile properties augment concrete's compressive qualities.

The advancement of bridge design ultimately relies on a material to perform safely, effectively and efficiently. With materials technologies and the development of composite materials constantly improving, the future potential in bridge design is almost boundless.

Enduring wood
The Chengyang Wind and Rain Bridge in Quangxi Province, China, was built in 1916. The elaborate deck with its five pavilions and numerous terraces is constructed of wood, the three piers are built in stone, and the roof is covered in ceramic tiles.

Stone

Stone is an ancient bridge-building material and the most enduring. Although it is weaker under tension than compression, stone's durability and strength make it particularly suited to bridges. Stone bridges take advantage of stone's compressive strength by arranging the material vertically (as piers) and as arches. Some bridges do rely on the tensile strength of stones, such as ancient or very small bridges where a slab is placed across a small stream or series of pillars traversing a river.

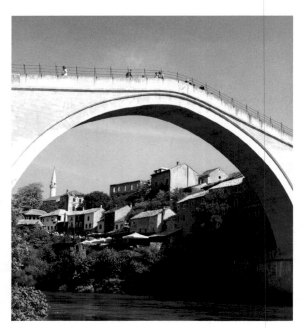

Bridging communities
The old stone bridge (Stari Most), in Mostar, Bosnia and Herzegovina, a 4 m (13 ft) wide and 20 m (66 ft) high humpbacked arch bridge, was built in 1566 by Mimar Hajruddin, a pupil of one of the Ottoman Empire's most famous architects, Sinan. It bridged communities on either side of the Neretva River for centuries before its destruction in 1993 during the Croat-Bosniak War. It was faithfully reconstructed and reopened in 2004.

Natural stone arches (right)

The oldest stone bridges occur naturally as eroded arches on coastlines and exposed areas of bedrock around the world. In the US state of Utah, the Arches National Park contains more than two thousand of these natural wonders, the most famous of which is the 16 m (52 ft) high Delicate Arch. These natural formations would have inspired ancient civilisations and demonstrated the capability of stone.

Stone ornament

The sculptural quality of stone is exploited in ornamental bridges, such as the Roman triumphal bridges, which performed important decorative or ceremonial functions and were furnished with elaborately carved stonework.

Stone beams (left)

The simplest stone bridges use stone as a beam, relying on its tensile qualities. As the distance between the supports at either end increases, so must the thickness of the stone. Although simple in theory, the technical sophistication of stone-beam bridges varies from the Chinese bridges where individual stones can weigh more than 200 tonnes to the diminutive 'clapper bridges' in the United Kingdom.

Timber

Flying eaves

The Chengyang
Wind and Rain
Bridge in China's
Quangxi Province
was built entirely
of wood in 1916.
It uses a method
of construction known
as flying eaves,
whereby layers of
overlapping beams
support a load in
the same manner
as in the distinctive
traditional Chinese
roof. Multi-storeyed
pavilions stand on
the five stone piers
supporting each
section of this
64 m (210 ft) long
and 3 m (10 ft)
wide, covered
pedestrian bridge.

Timber is an ideal material for bridge building – it is abundant, lightweight, comparatively cheap, easy to work, and extremely versatile. Consequently, timber bridges are found all over the world and come in all shapes and sizes, from the ornamental elegance of dainty footbridges to the utilitarian solidity of massive railway bridges. Timber's structural dexterity means that it can be used to carry loads in compression or in tension, and also enables it to be easily crafted into intricate shapes and designs.

Wooden trusses (right)
America's first covered wooden bridge was built across the Schuylkill River, Pennsylvania, in 1805. Depicting the bridge before it was clad, this engraving reveals how the deck is fabricated from wooden trusses with arched profiles that span between two stone piers.

Wooden arches
The five slender arches (each approximately 35 m/115 ft long and 5 m/16 ft wide) of the Kintaikyo Bridge that leap across the Nishiki River in Japan were originally built in 1673. The complex structure, originally built entirely of wood, is designed to resist heavy flooding. It was last rebuilt in 1953.

Ancient traditions
The 98 m (322 ft) long Wan An Bridge in Fujian, China, built more than a thousand years ago, is testament to wood's potential endurance. The region is famous for its covered wooden footbridges, each of which is constructed from segmental arches made up of wooden beams supported on stone piers.

Organic Material

Modern vines
Today there are three bridges across the remote Iya Valley of southern Japan measuring up to 45 m (147 ft) long and 2 m (6½ ft) wide.

One of the earliest methods of bridging large rivers was by felling a tree across the banks. This had its limitations – not least if a gorge was too wide or if there were no suitable trees. In some regions, nature offers assistance in other ways, such as with vines and vegetation hanging from the forest canopy. Over time, more sophisticated ways of manipulating organic materials other than wood into bridges have been invented.

Vines

For centuries, the only means of crossing the Iya Valley was the vine bridges, or *kazurabashi*, suspended 14 m (46 ft) above the river. These were first constructed in the 12th century so that retreating warriors could easily cut the vines to prevent their enemy's advance. Each one is regularly reconstructed but all are now reinforced with steel wires.

Roots (left)

In the drenched jungles of northeast India, the locals have devised a unique way of constructing bridges from a native fig tree, *Ficus elastica*. Rather than cutting the tree down, they manipulate the aerial roots growing from the tree's branches and train them across the river to create the suspension cables of living bridges that grow stronger with age. The decks are paved with stones and sometimes built over two storeys, creating double-decker suspension bridges.

Bamboo

Bamboo has been used in Asia for bridge construction for thousands of years. Its high compressive and tensile strength, flexibility, and lightness make it ideal for floating or suspension bridges carrying relatively small loads. The longest bamboo suspension bridges, up to 200 m (655 ft) long, were constructed in China's Sichuan Province. As a processed material, bamboo is now being used to build bridges capable of carrying vehicular traffic.

Brick

Ancient brick

The first sun-dried bricks were manufactured in Mesopotamia six thousand years ago. In the 17th century, their modern equivalent, kiln-fired bricks, were used to construct the 33-arched Si-o-se Pol Bridge in nearby modern-day Iran, one of the oldest and longest brick bridges in the world. Constructed over the Zayandeh River in the city of Isfahan, the 295 m (965 ft) long bridge has a central stone deck for vehicles which is flanked by covered pedestrian walkways separated by walls of 66 arches (two above each lower arch).

Bricks have been fabricated for thousands of years and are one of the earliest examples of a manufactured material used in bridge construction. In warm climates, the earliest bricks were made from uniform blocks of mud, clay or similar base materials and left to dry. The manufacture of weatherproof bricks relies on specialist materials and the expertise to burn bricks in kilns, which was first recorded in China's ancient capital, Xian, nearly four thousand years ago. Centuries later the Romans used similar techniques which, following their downfall, were lost in northern Europe until the Late Middle Ages.

Massive structures

The Goeltzschtalbruecke or Göltzsch Viaduct of southeastern Germany, is one of the most outstanding articulations of brick's versatility. Stretching over 500 m (1,640 ft) and nearly 100 m (330 ft) tall at its highest point, the multi-arched viaduct opened in 1851 and continues to carry modern high-speed trains.

Reconstructed brick

The ancient 22-arch brick bridge of Pul-i-Malan outside the city of Herat in Afghanistan is believed to date from the 12th century. Having survived many floods its bricks were lauded in local legends, but they succumbed to modern warfare and the bridge has recently been substantially reconstructed.

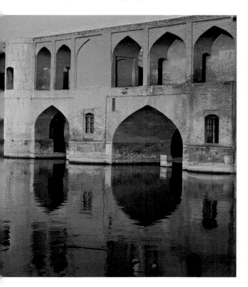

Brick styling

Brick was a favoured material of the Victorian era in Britain, and few architects used it more splendidly than Sir George Gilbert Scott, who designed the Clifton Hampden Bridge in Oxfordshire (1867). Scott's preference for the Gothic style of architecture, exemplified by his design for the Midland Hotel that fronts London's St Pancras Station, appears in the bridge's six pointed arches.

MATERIALS **Iron**

The Chinese were the first to use iron in bridges, fabricating iron links to create suspension chains. Local historians documented them in the 15th century and subsequent European travellers, including Marco Polo, marvelled at the sophistication of these pioneering suspension bridges. No other application of iron in bridge building matched these structures until the late 18th century, when the world's first cast iron bridge was constructed in Coalbrookdale, England. Throughout the Industrial Revolution iron was a popular construction material until it was superseded by steel.

Cast iron (right)

Iron transformed bridge construction during the 19th century. Opened in 1805, the 304 m (997 ft) long, 3.4 m (11 ft) wide and 38 m (125 ft) high Pontcysyllte Aqueduct was designed by Victorian engineer Thomas Telford. The trough, made from cast iron plates, is supported on 18 arches each formed of four cast iron ribs, and reinforcing outer plates bridge each of the masonry piers.

Iron Bridge (left)

No iron bridge is more famous than the one built across the River Severn in England, whose very name, Iron Bridge, derives from the material. The revolutionary bridge, which was a resounding advertisement for iron, was designed by Thomas Farnolls Pritchard using 378 tonnes of iron produced continuously over three months from nearby furnaces. Each huge cast iron member was transported to the site along the river and then hoisted into place to form an arched humpbacked bridge that opened on the first day of 1781.

Decorative iron

Industrialisation made it easier and cheaper to manufacture identical structural members that were not only structurally effective but could also be highly decorative. Elaborate shapes and patterns could be created during the casting process, and swiftly became a feature of iron and steel bridges.

Steel

Bridging the Tyne
More than 7,000
tonnes of steel were
used to construct
the Tyne Bridge
linking the cities
of Newcastle and
Gateshead. It was
designed by Mott,
Hay and Anderson.
The steel arch,
fabricated from
flat sheets of steel
riveted together
from which the
road deck is hung,
has a 161 m
(528 ft) span.
It was the longest
single-span bridge
in Britain when
opened by King
George V in 1928.

The advent of modern steel making marked a significant milestone in the development of bridge building in the late 19th century. Steel's superior strength in tension and compression compared with iron advanced the performance and construction techniques of metal bridges. Extruded steel wire was used to build the first modern suspension bridges, such as the Brooklyn Bridge in 1883, while other types of record-breaking bridges, including the 1 km (⅗ mile) long Forth Rail Bridge, the world's longest cantilevered bridge, were built from thousands of individual members joined together on site.

Steel arch

With a span of more than 500 m (1,640 ft) and a road deck 50 m (165 ft) above the waters of Kill Van Kull, New Jersey, the Bayonne Bridge, completed in 1931, was the largest steel arch bridge in the world until 1978. The arch is made up of 40 straight steel segments. Unlike similar bridges, it does not have decorative stone abutments.

Steel truss

The Brinnington Rail Bridge in Manchester, England, is a modern steel truss bridge that demonstrates the enduring quality of steel as a popular bridge-building material. The individual truss sections are riveted to prefabricated steel girders that form the upper chord and deck of the bridge.

Concrete

Prefabrication

An advantage of prestressed concrete is that it can reduce building times and costs by being prefabricated in sections and delivered to the construction site. By using prestressed concrete, the new Victory Bridge (2005) in New Jersey, whose 134 m (440 ft) main span made it the longest precast cantilever bridge in the United States, saved more than a year's construction time and millions of dollars.

In 1849, a French gardener, Joseph Monier, stumbled upon a radical invention when using iron mesh in concrete to make garden pots. By combining the tensile strength of iron and the compressive strength of concrete, ferroconcrete, commonly known as reinforced concrete, has since revolutionised the construction industry. Reinforced concrete has limited deflection and cracking characteristics and so prestressed concrete was invented in the early 20th century to enhance performance, thus enabling longer spans without increasing the depth of each section and, consequently, its weight.

Post-tensioned concrete (right)

Post-tensioning is similar in principle to prestressing. It is another method of increasing concrete's spanning capabilities by inducing compression forces in the parts of the structure likely to develop tensile forces. Unlike prestressing, post-tensioning is carried out after the concrete has been formed and set. Steel tendons within casing sleeves are located in the member either before the concrete is poured in situ or, in the case of prefabricated elements, as the individual segments are lifted into place. They are then pulled tight, forcing the concrete into compression. The tendons are then locked off.

Slimline (below)

The structural qualities of prestressed concrete mean that large structures can be built using a comparatively small amount of material. The slender appearance of the 252 m (827 ft) span arch across the Wilde Gera Valley in Germany would not have been achievable using standard reinforced concrete.

Prestressed concrete

Prestressing concrete is a method of enhancing the material's spanning capabilities. The concrete is cast around pre-tensioned steel cables, called tendons. As the concrete cures, a bond is formed between the tendon and the concrete. After curing, the tendon ends are released, forcing the concrete into compression. The tendons are positioned where tensile forces develop, usually at the bottom portion of a simple span.

Reinforced

Unloaded

Loaded

Prestressed

Unloaded

Loaded

Glass

MATERIALS

Revolutionary glass
A footbridge over the Grand Union Canal, Paddington combines a 3.5 m (11½ ft) diameter glass tube with a 7.5 m (24½ ft) long stainless steel helix to create a functional bridge and a work of art. To allow boats to pass, the helix rotates and the bridge retracts.

Glass is among the most exhilarating materials to be used in bridge construction. Modern glass-making techniques have allowed glass to overcome many of the problems that until relatively recently prevented it from being used in bridges. Today, toughened glass can be manufactured in almost any shape and size. Glass is not only functional and safe, but it is also durable and resistant to heavy wear and weathering. Although glass has exceptional compressive strength, most applications in bridge design rely on other materials for structural integrity, such as steel frames. This will almost certainly change as the performance of glass improves, allowing bridges to be built entirely of glass.

Glass floor

20 m (66 ft)
projection

1,200 m
(3,950 ft)
drop

Transparency (above and right)

The obvious advantage of glass is its transparency. Few things match the thrill of being able to enjoy unobstructed views of scenery beneath your feet. Reinforced glass floors have been installed in high-rise structures for years, but nowhere in the world compares with the glass floor of the cantilevered horseshoe bridge over the Grand Canyon, USA. With a drop of more than 1,200 m (3,950 ft), the glass and steel bridge creates the highest view from a man-made structure in the world.

Interior design

Glass bridges are becoming a common feature of modern interior design, particularly in commercial or office spaces, and even in some public buildings. In 1997, a glass bridge suspended on 1 mm ($\frac{1}{32}$ in) thick steel wires was installed in the Challenge Materials gallery of London's Science Museum.

BRIDGE TYPES Introduction

Irrespective of age or materials, every bridge can be categorised according to its type. The four basic types of bridge are beam, arch, cantilever and suspension. The sequence of these types also reflects an approximate chronology of the development of bridge building. The most basic bridge type is the beam, which can be as simple as a log across a stream. Building an arch requires an element of constructional acumen, and a cantilever relies on a more developed appreciation of engineering. Although primitive examples of suspension bridges exist, modern suspension bridges offer the largest spans and are often the most challenging to build.

Other types of bridge are variations on these four basic examples. Cable-stayed bridges, for example, are a development of suspension and cantilever bridges, whereby each cable is attached to the tower and fixed directly to the deck rather than carrying the loads from the deck through hangers to two main cables hung over the towers.

Truss and moving bridges on the other hand can employ any one (or a combination) of the four main types in a structural configuration that is either constructed from truss sections or is movable. Many bridges are designed using a combination of types and such bridges are frequently referred to as hybrid.

Old and new
The Forth Road Bridge (1964) was the first bridge in Europe to use the modern method of cable-spinning during construction, using more than 30,000 km (18,640 miles) of wire. Behind is the cantilevered Forth Rail Bridge (1890).

Beam Bridges

Box girders
One variation of the beam is the box girder, which gains its strength from a hollow rectangular cross section. Prestressed concrete box girders are used in the construction of the 13 km (8 mile) long and 72 m (236 ft) wide Rio–Niterói Bridge (1974) in Brazil.

In its simplest form a beam bridge is a horizontal deck supported at both ends, but there are many different types. A beam bridge has to withstand vertical loads that generate vertical shear forces, horizontal tension and compression forces, and when only loaded on one side, torsional twisting forces over its length. The vertical shear forces are shared between the supports. A beam's spanning capability can be intuitively characterised by the ratio between the depth of its cross section and the distance between the supports (span-to-depth ratio) – if the distance between the supports at either end is too great, the beam may develop forces that it is not strong enough to withstand.

Basic beams

The simplest illustration of the beam bridge, and possibly the earliest example used by man, is the fallen tree across a river. Here, the beam comprises the tree trunk, which is supported at both ends by the river banks.

Continuous beam bridge

Pier

Deck

Continuous beam (above)

An alternative to the simply supported span bridge is the continuous beam, which uses a single beam supported over numerous supports, or piers. The advantage of a continuous beam is that depths can be reduced as the beam works more efficiently by developing tension forces in the top portion of its cross section over the piers.

Simply supported beams (below)

Bridges that comprise individual beams between supports are known as simply supported span bridges. This means that although there may be repetition (numerous supports and numerous spans) the beam is not a continuous member over the top of the supports. Simply supported beams are used where the overall span requires only one length of beam or multiple spans where movement joints are required.

Simply supported beam bridge

Pier

Joints *Deck*

Joints *Pier base*

Beam Bridges

Stone beams

Before modern advances in engineering and materials, beam bridges spanning large distances used a large number of beams supported on closely spaced piers.

Luoyang Bridge in Quanzhou, China, built in 1060 CE, is more than 1 km (⅗ mile) long. Its 46 masonry piers support huge carved granite stones.

Spreading the load (left)

Spreading the load across a series of beams arranged in parallel, as illustrated in this conventional highway bridge, helps to overcome twisting that could occur when weight is unevenly distributed on the deck. Here the load on the deck is supported by four evenly spaced steel girders connected at each pier, which transfers the load into a single column, minimising the bridge's impact at ground level.

Luoyang bridge beams (right)

With some beams as long as 20 m (66 ft), the thickness they need to be to support even their own weight ensures each one weighs up to 200 tonnes – an extremely inefficient ratio. Because eight beams are required for each 5 m (16 ft) wide span, the Luoyang Bridge was constructed with more than 350 of these massive stone beams linking 46 streamlined masonry piers.

Mega beams (left)

Beams are ideally suited to extremely long bridges, since the basic principle of a beam supported by piers at either end can be replicated infinitely. It is no surprise that the world's longest bridge – the Danyang–Kunshan Grand Bridge, 165 km (103 miles) long, completed in 2010 – is a beam bridge. Hundreds of prestressed concrete beams have been laid across multiple piers.

Arch Bridges

Line of forces

In an arch bridge, applied vertical forces are carried through the arch in compression and into the abutments, which resist these forces with vertical and horizontal reactions. In traditional stone arches, each block is called a voussoir and the stone at the apex is called the keystone. This locks the arch in place and ensures that vertical forces are translated into lateral lines of force.

The arch is naturally a very strong structure. Its occurrence in the natural environment predates and doubtlessly inspired human attempts to replicate it. These attempts were first in stone, but since then many other materials, including wood, brick, iron, steel and prestressed concrete have been used. An arch works by transferring the vertical load through its curve to the supports at either end, known as abutments, where the load is dissipated into the ground. There are several ways that bridges exploit the arch – some place the deck on top of the arch, some suspend the deck beneath the arch, while others allow the deck to pass through the arch.

Abutments

Vertical force

Keystone

Line of force

Voussoir

Footers

Ancient arches

The simplest type of arch can be found in ancient stone structures such as tombs, where two colossal stones placed diagonally against one another transfer the loads either side of the opening beneath.

Half-through arches (below)

The Lupu Bridge (2003) is a box-steel arched bridge crossing the Huangpu River in Shanghai, China. With a span of 550 m (1,800 ft), it is the second longest arched bridge in the world. The bridge comprises two massive steel arches braced together and from which the deck of the bridge hangs on steel wires. Because the deck intersects the arch above the abutments, it is known as a half-through arch.

Deck-arch bridge

Thrust forces resisted at abutments

Through-arch bridge

Tensile tie forces in deck structure

Half-through-arch bridge

Deck may be utilised to tie arch, reducing lateral thrust forces at the foundations

Arch types (left)

Single-span arched bridges traditionally caused traffic to climb up and over their humps. Nowadays, different types of arched bridge maintain a horizontal deck, including half-through arches, deck arches and through arches, where the hanging deck intersects the arch at the abutment (see also pages 110–111).

Steel arch (right)

Modern construction materials and techniques have allowed arch bridges to become increasingly slender. The lean lines of the half-through arch bridge (1990) over the Roosevelt Lake in Arizona, USA, is an example of how unobtrusive modern bridges can be when constructed in areas of outstanding natural beauty.

Open
spandrel

Spandrels (above)

The inherent weight (dead weight) of any bridge can cause it to collapse, so reducing weight is often a key consideration in design. Creating holes in the space between the extrados, the exterior curve of an arch, and the abutment not only reduces the weight of the bridge, but also allows flood water to pass through, reducing the bridge's resistance to water. These practical voids are called open spandrels, as seen above in the Pontypridd Bridge in southern Wales.

Deck arch (above)

An arch that supports the bridge deck above it is called a deck-arch bridge. The Eads Bridge over the Mississippi in St Louis, USA, which carries road and rail traffic, was the longest deck-arch bridge in the world when it opened in 1874. The load on a deck-arch bridge is supported by the arches beneath it and, in this case, also by the piers in between.

Truss Bridges

The classic truss
Many railway bridges adhere to the classic image of a truss bridge. The Colorado River Bridge, USA, illustrates the truss's structural geometry, showing the triangular sections formed by the diagonal members between the vertical columns and the horizontal chords.

Truss bridges are one of the earliest forms of modern bridge design. There are many varieties of truss, but as a structure they all take advantage of the inherent strength of the triangle. The straight members in a truss are subject to compression, tension, or a combination of both (not at the same time) depending on the dynamic forces applied to the truss. Compression members are typically larger in cross section to withstand buckling effects. The truss is a very efficient structure as it uses comparatively little material and has a very good strength-to-weight ratio. Trusses can be used in all types of structures, including beams, arches and cantilevers.

Wooden truss

Low Pratt truss

Wooden truss

The earliest truss bridges were constructed in wood. This design by the Italian Renaissance architect Andrea Palladio (1508–80) illustrates the truss's triangular geometry. The bridge is divided into three sections, with two inclining triangular approaches supported at one end by abutments and at the other end supporting the bridge's horizontal central section.

Low Pratt truss

This particular configuration, in which the diagonals slope down into the middle of the bridge, is called the Low Pratt truss and was invented by Thomas and Caleb Pratt in 1844.

Continuous truss

Some truss bridges are built using a repeated series of vertical, horizontal and diagonal members to create a continuous structural framework. The Great Crumlin Viaduct (1857), in southern Wales, used continuous trusses in both the construction of the bridge deck and the supporting pylons. It was dismantled in 1967.

Truss Bridges

Vertical lift bridge

Supporting pylons

Deck

Movable section

Moving loads (above)

The truss is well suited to moving bridges since it has a good resistance to moving loads. In the case of a lifting bridge, where heavy loads move across the bridge's span, the truss is capable of supporting the load at any position along its length, transferring it into the supporting pylons on either side.

Hybrid trusses (below)

Often a variety of truss types are used in a single bridge. The original Kentucky and Indiana Bridge (1886) across the Ohio River used a number of different truss types, including arch, beam and cantilever trusses constructed from wood and steel throughout its 1 km (³/₅ mile) long span.

Lenticular truss

So-called because of the distinctive lens-shape created by the upper and lower chords, the lenticular truss was used by Isambard Kingdom Brunel in his design of the Royal Albert Bridge (1859). Each truss comprises an upper chord of tubular iron in compression and two lower chords of iron chains in tension, which are cross-braced to withstand wind load. The railway deck is suspended beneath. Unlike an arch, which transfers horizontal forces into the piers, the lenticular truss is simply supported on the piers.

Lattice truss

The lattice truss uses a series of diagonal members laid in a lattice between larger and stronger horizontal members. This example shows how a truss can also be read as a beam, with each lattice truss section forming a large beam supported on piers. Notice how the diagonals get larger as they carry the load back to the supports.

Moving Bridges

Rolling bascule

The most common form of moving bridge is the bascule bridge, which works on the same principle as a seesaw. The famous Pegasus Bridge (1934 and 1994) in Normandy, is a rolling bascule – rather than opening about a pivot point on the deck, the structure rolls open.

Although bridges are designed to ease passage they can also be an obstruction – or worse. A road bridge over water might be good for cars yet restrict boats, and a castle drawbridge that failed to withdraw would have been a matter of life and death. Moving bridges are designed to overcome the problems associated with the permanence of a bridge. There are a variety of moving bridges to suit all sorts of situations, from bridges that rise into the air to ones that sink underwater; and from those that tilt to those that curl. As with all types of bridge, the design of moving bridges has evolved as new materials have become available.

Fully open

Curling

Temporary bridges are often used in military situations, where advancing or retreating soldiers can prevent their enemy following them. They are also lightweight and easy to move, like this 16th-century example that can be unfurled across a river or curled up to transport.

Closing

Closed position

Drawbridge

Among the most renowned moving bridges is the drawbridge, made famous by medieval European castles. The drawbridge opens and closes by means of retractable ropes or metal chains attached to the farthest ends of the bridge's deck. Counterweights would often be used to make the process easier.

Bascule

The bridge deck, or 'leaf', of a bascule bridge is raised using a counterweight, making the bascule energy-efficient and swift. Bascule bridges can have either one leaf or two. A bascule bridge with two movable sections that meet in the middle is known as a double-leaf bascule. Lowering the counterweight located on the top of the truss section or on a concealed back span raises the leaf of the bridge.

Moving Bridges

Bridge deck lowered   Bridge deck raised

Tilt (above)

Tilt bridges rotate at either end but for this action to have a useful purpose the bridge deck has to be curved, restricting its use to walking or cycling. In the case of the Millennium Bridge at Gateshead, the arch that supports the deck in its closed position rotates down as the deck lifts up, acting as a counterbalance. As the deck lifts it is able to hold itself up without the arch.

Vertical lift (below)

Bridges that raise the central deck of the bridge vertically so that it remains parallel with the approaches are called vertical-lift bridges. The deck is raised either on two towers at either end of the span or by hydraulic jacks underneath the span. Vertical-lift bridges have a better weight-to-strength ratio than bascule bridges, so they can carrier heavier loads.

Swing

Swing bridges rotate from a central pivot in the middle of the span. Because they do not require a counterweight they can be comparatively light, but the central pier from which they pivot can be a disadvantage and an obstruction in narrower channels.

Folding

Another means of retracting bridges is to fold them, like a concertina. Because of the comparatively fragile nature of moving parts, this method is suited to smaller bridges where each section is not too heavy and the moving parts are not exposed to too much stress.

Transporter

Bridges that carry a section of deck across the span are known as transporter bridges. Most operate by suspending the deck on cables from an elevated structure supported on two piers at either end of the bridge. The deck, known as the gondola, moves from one side to the other using a mechanised system of steel cables and pulleys.

Submersible

The submersible bridge is a rare type of movable bridge that lowers into the water to allow boats to pass over the top. More of the boat sits above water than below, so a bridge that sinks into the water does not have to travel as far as one that rises.

Cantilever Bridges

Arch or cantilever?
Cantilever bridges can be mistaken for arched bridges because the two cantilever arms often form an arc as they do here at Wandsworth Bridge, London. The bridge comprises two balanced cantilever sections standing on piers. The movement joints in the centre of the span are just visible.

A cantilever is a structure projecting in one direction from an upright support. If the structure extends in both directions to form a 'T' shape, it becomes a balanced-cantilever. In bridge design, the two parts of the balanced structure are known as the cantilever arm and the anchor arm. Many different types of cantilever are used in bridge construction and they can often be difficult to identify, not least because they can masquerade as arches due to the arced shape of some cantilever arms. The easiest way to read a cantilever from an arch is to look for the joint between two cantilevers, which will be at or near the centre of the span, whereas the joint between two arches will be at the pier.

Niagara cantilever bridge during construction

Completed bridge

Niagara cantilever construction

Constructing cantilever bridges is relatively straightforward because with the anchor arm tied back, the cantilever arm can be built incrementally out from the pier to form the main span. As a method it is ideal to use over deep gorges or busy rivers as the bridge requires no temporary support from below. These illustrations of construction and design of the bridge at Niagara Falls show the process of extending the cantilever arm out from the pier. The finished bridge comprises a pair of balanced cantilevers supporting a deck on top.

Lower floor beam

Upper lateral cross-bracing

Suspended span

Cantilever arm

Lower lateral

Anchor arm

Reading the cantilever

Each section of a cantilever bridge can usually be read as three distinct parts: the anchor arm, the cantilever arm, and the suspended span. The pier fixed to the anchor arm resists the uplift forces. The pier on which the cantilevers are balanced resists the compressive forces. The top chords are in tension. The bottom chords are in compression and the diagonal members carry the load back to the support.

53

Cantilever Bridges

Oakland Bay Bridge

The east span of the double-decker Oakland Bay Bridge (1936), which is also often known as the Bay Bridge, crosses San Francisco Bay, California, USA. The bridge comprises a series of small, medium and large truss spans, the largest being 430 m (1,411 ft) long and is formed by two large balanced cantilevers. The bridge is due to be replaced in 2013.

Deck   Arch   Anchor arm

Pier

Balanced cantilever

The appearance of the arch formed on the underside of the two cantilever arms is misleading, as the bridge does not function as an arch. Instead of transferring loads through the arch to abutments, the loads in a balanced cantilever structure are transferred to the piers. The downward forces in the main span are countered by the anchor arm, which is fixed securely to the piers of the approach road.

Extending the span

The central span of cantilever bridges is often lengthened by inserting a suspended section between each cantilever arm. In this classic photograph, the person seated in the middle represents the suspended span supported by the cantilevers either side. The wooden posts represent the lower chords in compression and the men's arms represent the upper chords in tension. Both sets of chords are anchored at either end to prevent the cantilevers toppling inwards.

Cantilever truss span

Truss cantilever

Cantilever bridges are commonly truss bridges. Making the structure stronger over the supports causes it to behave as a cantilever with the central section acting as a suspended beam. Placing two balanced cantilevers next to one another creates a large span between their supporting piers. The Oakland Bay Bridge is a good example of this method of cantilever bridge design, in which the cantilever arm extends out to form half of the main span and is secured by the anchor arm, which is fixed to piers.

Suspension Bridges

The cable

The main cables on modern suspension bridges are constructed from a single cable spun back and forth around the entire structure. A 30,000 km (18,640 mile) strand was used to make the suspension cables for the Forth Road Bridge (1964) in Scotland.

Suspension bridges have existed for thousands of years, but they have be perfected in the last two centuries thanks to the advancement of modern materials and construction practices. A suspension bridge relies on the tensile strength of a cable, from which a deck is hung. The earliest bridges to exploit this idea were footbridges suspended from ropes or vines strung across a stream, creating a catenary, or natural, arc. The steel cables used in the construction of modern suspension bridges are anchored to the ground and typically spun in situ over two pylons to create the main span between. The vertical cables connecting the main cables to the deck are called hangers.

Structure

The construction of a suspension bridge begins with the pylons. The main cables are created by spinning a continuous steel wire back and forth over the pylons and through loops at the anchorage. Once complete, these cables are bundled together to create one large cable. The deck beams are then hung on smaller cables. The decks of many modern suspension bridges are box girders. These beams are connected and strengthened to resist vertical loads as well as lateral loads caused by wind. Methods of strengthening vary depending on the type and span of the bridge, but can include truss beams, cross-bracing or guy wires.

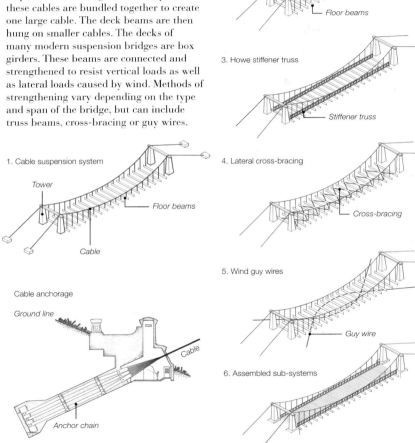

1. Cable suspension system

Tower

Floor beams

Cable

Cable anchorage

Ground line

Cable

Anchor chain

2. Timber deck

Floor planks

Stringers

Floor beams

3. Howe stiffener truss

Stiffener truss

4. Lateral cross-bracing

Cross-bracing

5. Wind guy wires

Guy wire

6. Assembled sub-systems

Suspension Bridges

Iron and steel

The suspending elements of early suspension bridges were made from iron chains or a series of interconnected flat bars called eyebars. A more effective method of using wire cables was devised in the early 19th century. Cables are stronger and possess more redundancy – a worn wire in a bundle of thousands is not catastrophic, but a fractured eyebar could be. Permanent wire-cable bridges were first built in the early 19th century, but the development of steel swiftly produced the first steel-cable suspension bridge: Roebling's Brooklyn Bridge (1883), New York.

Efficiency

Suspension bridges are the most efficient means of spanning large distances. They can be constructed without obstructing traffic beneath and require no falsework. They also use comparatively little material while being able to carry heavy loads.

Brooklyn Bridge

The steel wires were spun continuously over the pylons creating four main suspension cables that are anchored into the bedrock of Manhattan and Brooklyn. The Brooklyn Bridge's main span of 486 m (1,594 ft) made it the longest suspension bridge in the world for two decades.

Failure

Suspension bridges are very effective at withstanding vertical loads, but their slender lines and flexible components expose them to dynamic loads that exceed normal carrying loads, such as cross winds. These loads, if frequent and not countered in the deck design, can be catastrophic, as with the collapse of the Tacoma Narrows Suspension Bridge in 1940.

Cable-Stayed Bridges

Versatility

Cable-stayed bridges were first used for situations that were too large for cantilevered bridges and too small to warrant a suspension bridge. However, with improvements in design and construction cable-stayed bridges are now rivalling suspension bridges, with spans exceeding 1 km (⅗ mile), making them a very versatile bridging solution. The main span in the middle of the cable-stayed Oresund Bridge (2000) that carries railway and road traffic between Sweden and Denmark is 491 m (1,611 ft) wide and supported by four independent needle towers.

Cable-stayed bridges appear similar to suspension bridges, but are in fact more like the cantilever bridge. There are many different types of cable-stayed bridge, but they are characterised by a pylon, or tower, supporting the deck with a series of individual cables in tension. In all cases the deck is built out from the tower in a series of cantilevers. These are arranged on both sides of the tower, which balances the structure and requires no anchorage. Because the deck is being pulled back to the tower and is therefore in compression, it has to be a very rigid structure; unlike the suspension bridge, in which the light deck merely hangs from the suspension cables.

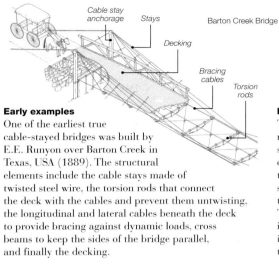

Barton Creek Bridge

Cable stay anchorage
Stays
Decking
Bracing cables
Torsion rods

Early examples

One of the earliest true cable-stayed bridges was built by E.E. Runyon over Barton Creek in Texas, USA (1889). The structural elements include the cable stays made of twisted steel wire, the torsion rods that connect the deck with the cables and prevent them untwisting, the longitudinal and lateral cables beneath the deck to provide bracing against dynamic loads, cross beams to keep the sides of the bridge parallel, and finally the decking.

Harp and fan (below)

There are two principal methods of arranging the supporting cables in typical cable-stayed bridges. One is the fan design, in which the supporting cables are gathered towards the top of the tower. The other is the harp design, in which they are arranged in parallel as they rise to the tower.

Fan design

Stays

Cable-stay anchorage

Cable-stay anchorage

Harp design

Stays

Cable-Stayed Bridges

Unlimited length

The main spans of modern cable-stayed bridges cannot match those of the longest suspension bridges, but the advantage of cable-stayed bridges is that they can repeat continuously. Because anchor points for cable-stayed bridges lie in the towers, the bridges can be extended by multiplying the towers. The Ting Kau Bridge (1998) in Hong Kong uses a series of towers to traverse a large expanse of water. Extra-long cables, measuring 465 m (1,525 ft), increase its longitudinal stability.

Construction (right)

The construction of large, modern, cable-stayed bridges begins with the approach roads and tower. Once these elements are in place, the deck is hoisted into position in sections hung from separate cables. When complete the whole deck is connected and stiffened. Construction is comparatively straightforward and has two main advantages: it requires little or no falsework and there is limited obstruction to shipping beneath.

1. Erection of piers and support spans

Span

Central tower

2. Construction of work station on central tower

Erection derrick

3. Installation of temporary stay cables and first erection cables

Erection cable

Temporary backstay cable

Backstay anchorage

4. Extension of central span

Temporary backstay cable

Erection cable

Temporary backstay cable

5. Completion of central span and removal of temporary cables

Erection cable

Hybrid Bridges

Arch and trusses

The Dom Luís I Bridge (1886) over the Douro River, Portugal, comprises an arch supporting a railway deck with a lower road deck hanging beneath. The upper decks are steel truss beams supported on pylons, with only the central portion being carried by the arch.

Relatively few bridges are true examples of a single bridge type. Most are hybrids of some sort, but some are more explicitly hybrid than others. Hybrid bridges are the most challenging bridges to read, because two or more structural types are working in tandem. Certain bridges are designed as hybrids, whereas others become hybrid throughout their working lives, as they are strengthened or altered to meet changing circumstances. The inherently varied character of hybrid bridges means that there is an infinite variety of types. The following are just a selection of examples to illustrate how different bridge types can be used in a single bridge.

Cables and box girders

The decks of cable-stayed bridges are an important structural component of the bridge. They have to withstand the compressive forces imposed by the cables pulling them back towards the towers and the bending forces imposed by the loads pushing them downwards. The Erskine Bridge (1971) in Scotland uses cable stays to support the deck, which is constructed from welded segmental box girders.

Unintentional hybrids

Some bridges are not designed to be hybrids but become so during their lifetime. The Albert Bridge (1873) in London was a originally a cable-stayed bridge with 32 iron rods supporting the deck. Problems with the bridge led to steel chains, similar to those of a suspension bridge, being added ten years later. In the 1970s a concrete pier was inserted in the middle of the reinforced span to reduce the load on the towers.

Arches and beams

The engineer Robert Stephenson inserted arches into the wrought iron ribs to create the individual spans of the High Level Bridge (1849) in Newcastle, England. The distinctive section combining a tied arch with horizontal and vertical ribs, sometimes referred to as a bowstring girder, is extremely rigid and was designed to withstand the loads imposed by this railway and road bridge.

Extradosed bridges

By combining the attributes of cable stays and beams, some bridges are able to achieve greater spans and narrower decks than beam bridges and not incur the costs associated with large cable-stayed bridges. Bridges that combine the two are referred to as extradosed and are very distinctive.

The first section of the deck is directly supported on the tower to act as a continuous beam. The cable stays support the next section of deck, while the central portion of the span is suspended by outer cables. The cable stays of the Ganter Bridge in Switzerland (1980) are embedded in the prestressed concrete arm.

Golden Ears Bridge (above)

The longest extradosed bridge in North America is the Golden Ears Bridge (2009) in Canada, with four towers spanning over 2 km (1¼ miles).

Bridge series (below, left)

Some modern bridges are so large that they are a series of bridges rather than a single one. The 42 km (26 mile) long Qingdao Haiwan Bridge (2011) in the bay of the Chinese port of Qingdao is the world's longest bridge over water. It comprises cable-stayed sections, and both prestressed and post-tensioned concrete beams.

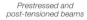

Cable-stayed section

Prestressed and post-tensioned beams

Sunniberg Bridge (right)

Another advantage of the extradosed bridge is that its relatively short spans can create a curved deck. The Sunniberg Bridge (1998) in Switzerland is a reinforced concrete extradosed bridge with four piers and five spans that curve in an arc through the steep valley.

Introduction

Bridges are designed to satisfy a wide range of different uses, from the obvious, such as vehicular, railway, cycling and pedestrian, to the more obscure, such as carrying water. Many are even designed to cater for multiple uses.

Most bridges are designed for a specific use (or uses), but the actual use often changes to adapt to the shifting needs of different users. Bridges often acquire a use for which they were never originally intended, which can evolve much later. Ancient bridges that were designed to carry battalions of soldiers or essential water supplies now

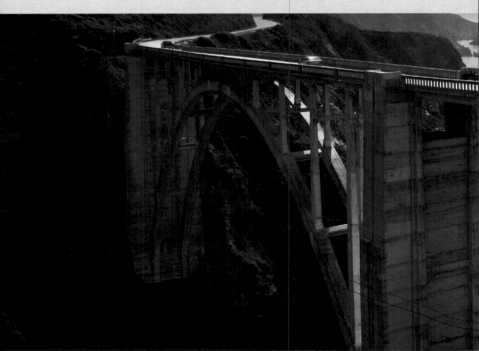

bear only the weight of tourists and ramblers. Others are victims of their own success and have to be widened or converted from a single deck to a double deck.

Large bridges often have multiple uses. Since they are designed to carry the heaviest loads, they can also accommodate the relatively negligible weight of pedestrians and cyclists. Some bridges have become so iconic that the chance to walk over their structure or to jump from it while attached to a bungee rope becomes a pursuit in its own right.

Embracing the car
Thousands of bridges were built to facilitate the expansion of the road network in the United States. One of the most famous is the reinforced concrete deck-arched Bixby Creek Bridge (1932) on the West Coast.

Pedestrian

Linking spaces

The Millennium Footbridge (2000) connects St Paul's Cathedral in the City of London to the South Bank of the River Thames. The high-tech, horizontal suspension bridge has transformed the urban spaces on both sides.

The oldest bridges were designed to walk over. However, the primitive origins of the footbridge belie the range and depth of its sophistication today. In recent years, the pedestrian bridge has been enjoying a renaissance as urban planning and design begin to retreat from decades of motor-car dominance and embrace a more sensitive and humane approach that prioritises the pedestrian and cyclist. The power of the pedestrian bridge to enhance the urban landscape is being proven all over the world and is resulting in all manner of innovative and challenging solutions in bridge design.

Romance

Unencumbered by other forms of traffic, pedestrian bridges have the capacity to be places of tranquility, contemplation, romance, and sites from which the city or surroundings can be observed from exciting and unobstructed perspectives. Few bridges in the world are more renowned for their romantic possibilities than Venice's Ponte di Rialto (1591), designed by Antonio da Ponte.

Urban regeneration

The Merchant's Bridge (1996) was an integral component of the regeneration of the Castlefield district of Manchester. The design brief demanded a striking and charismatic structure that would transform this formerly neglected part of the city. The solution, designed by Whitby & Bird, was a distinctive 85 tonne steel structure characterised by the sickle arch that supports the curved deck.

Design innovation

The comparatively small-scale and bespoke nature of pedestrian bridges offers exceptional opportunities for design innovation. The single span Passerelle Léopold-Sédar-Senghor (1999) in Paris is a steel arch bridge with a uniquely designed intersecting split-level deck allowing people to walk directly over the main arch or on its supported upper deck.

Water

Transport
Aqueducts have helped to boost trade since the 18th century and are still used for transporting raw materials. The world's longest navigable aqueduct is the Magdeburg Water Bridge (2003) in Germany, which is nearly 1 km (⅗ mile) long.

Bridges that carry water are called aqueducts – a name that reveals its most celebrated usage. By joining together the Latin words *aqua* ('water') and *ductus* ('to lead'), the Romans defined a type of bridge that has continued to be used for thousands of years. The earliest aqueducts were built for irrigation, but as cities grew and trade increased, water was needed for more than just growing crops. The Romans were expert builders of aqueducts and erected them to deliver clean water into cities for drinking and washing. Modern aqueducts are used for transportation and were critical during the Industrial Revolution in linking networks of canals.

Pont du Gard, Nîmes, France (above)

The Romans were the most prolific aqueduct builders, whose engineering skills were not surpassed for more than a thousand years. One of the most famous Roman aqueducts is the Pont du Gard in southern France. Once serving the city of Nîmes, the aqueduct's distinctive profile of three layers of arches stands 48 m (157 ft) high and 275 m (900 ft) long.

Aqua Claudia, Rome (below)

No Roman city consumed more water than Rome itself, with its thousands of baths and its position at the heart of the empire. To satisfy this demand, Emperor Caligula ordered the construction of the impressive stone arched Aqua Claudia in 38 CE, completed in 52 CE. This aqueduct supplied water to 14 districts of Rome.

Industry

Canals were essential to the Industrial Revolution. In designing the canal systems, engineers also had to design the aqueducts to bypass existing obstacles. The Stanley Ferry Aqueduct (1839) in Yorkshire carrying the Aire and Calder Navigation is the cast iron aqueduct with the longest span in the world.

Vehicular

Banpo Bridge

The Banpo Bridge (1982) across the Han River in Seoul is a major road bridge constructed over another road bridge, the Jamsu Bridge. To enhance this double-decker bridge, the city installed the Moonlight Rainbow Fountain along its length.

Up until the 19th century, most bridges had to carry little more than pedestrians and horse-drawn carts. With the advent of steam power and, later, the combustion engine, vehicular traffic increased exponentially and bridges became larger and stronger in order to bear the weight of modern traffic. Over the last century, the vast majority of bridges were designed to accommodate vehicular transport. Vehicular bridges come in all shapes and sizes. Some small structures provide improved links in established transport networks, while other larger ones create entirely new transport networks, carrying millions of vehicles every year across valleys, bays or even seas.

Spirit of freedom (right)

No country has embraced the motor car and the spirit of individual freedom that this means of transport represents more passionately than the United States. Bridge building was essential to the expansive road network as it spread westwards from the late 19th century. The Bixby Creek Bridge (1932) on America's West Coast is one of countless bridges built during the heyday of American highway construction.

Car boom

The 8.5 km (5¼ mile) long General Rafael Urdaneta Bridge (1962) in Venezuela, which combines a cable-stayed central section and prestressed concrete beam approaches, is an example of the enormous investments made in highway infrastructure over the last 50 years.

Rokko Island Bridge

The 217 m (712 ft) long, steel tied-arched Rokko Island Bridge (1993) in Kobe, Japan, forms part of a busy expressway and is another example of a double-decker road bridge designed to carry large volumes of urban traffic.

Railway

Glenfinnan Viaduct
The Glenfinnan Viaduct (1901) in Scotland was the first in the world to be built entirely from concrete. The structure, with its 21 semicircular arches, was built by Sir Robert McAlpine using non-reinforced concrete.

With the advent of the railways in the 18th century new methods of construction had to be devised that could carry these new unprecedented loads. The first railways did not carry passengers but heavier cargo, such as coal and other raw materials that fuelled the Industrial Revolution. The oldest surviving railway bridge is Causey Arch (1727) in County Durham, a stone arch erected by the mason Ralph Wood. Since these early beginnings, railway bridges have been built worldwide, using every method of construction, to carry heavier loads and cross ever wider distances.

Del Garda Viaduct

Viaducts are among the most celebrated and beautiful types of structure designed to carry trains. The pointed arch stone viaduct on the Del Garda Railway in Italy was built in 1852, but destroyed by air raids during the Second World War.

Garabit Viaduct

The Garabit Viaduct (1884) in France was built by one of the most famous engineers of the 19th century, Gustave Eiffel. Dominated by the 165 m (540 ft) steel truss arch that leaps across the Truyère River, the entire length of the railway bridge is more than 0.5 km (1/3 mile) long. The use of trusses was deliberate, to minimise wind resistance.

Bridge over the Kwai

Linking Thailand and Burma, the bridge over the Kwai Yai River was one of the greatest feats of engineering undertaken by the Allied prisoners who built the railway line during the Second World War. The bridge was immortalised in Pierre Boulle's novel *The Bridge on the River Kwai*, which was made into a feature film in 1957.

Military

The Bailey bridge
The Bailey bridge was designed in Britain during the Second World War. The success of the steel truss design was due to its transportability and ease of construction by hand and often under fire.

Bridges designed and constructed specifically for military purposes are a unique breed. Unlike other types of bridge they are temporary, since it is essential for them to be easily and quickly constructed and deconstructed, as well as to be transportable. Military bridges have been used for thousands of years and are often critical to strategic success. The earliest types of military bridge were of the pontoon variety, in which buoyant sections are bound together to form the deck. Today, with the weight and size of modern military hardware, military bridges are often a source of extraordinary innovation.

Modular design

The modular design of 3 m (10 ft) sections gains its strength from the lattice truss that forms the side of the bridge, each of which weighs 260 kg (573 lbs) and requires just six men to construct it. A testament to its enduring design, the Bailey bridge is still used today in peacetime and sometimes even as a permanent structure.

Specialist bridges (above)

Modern armies are equipped with a wide range of tools that can create or perform the function of bridges. This 3.5 m (11½ ft) wide and 20 m (66 ft) long bridge is made from a lightweight alloy and can be folded in half and carried on a converted tank.

Pontoon (below and right)

The pontoon is the oldest and the classic example of military bridge design. A series of floating sections made of boats, inflatables, or natural materials such as bamboo, supports a lightweight deck. Some form of cross-bracing is usually required to provide sufficient structural rigidity to prevent the bridge from being washed away by fast-flowing rivers.

Introduction

The majority of bridges are rarely noticed, let alone the talents of those who designed them. The names of the countless men and women who have designed and overseen the construction of bridges are invariably forgotten, but occasionally a bridge will make (or even break) a reputation and career. Through the centuries, great bridge designers have helped to transform both their trade and their society, advancing commerce and communications regionally or improving the urban realm locally.

Some of the most prolific and talented bridge designers have exploited the unprecedented developments in materials and construction technologies over the last two centuries. Conventionally, most of these often-celebrated figures have been engineers, but their ranks also include architects, artists and even sculptors. Common to each of these individuals is a talent for designing effective and efficient bridges that do not merely provide a connection between two points, but inspire and thrill the user or onlooker in the process.

The following is a small selection of the hundreds of influential bridge designers who have helped not only to transform and improve the world we live in, but who have inspired generations of fellow professionals and members of the public through the beauty of their work.

Victorian engineer
Few engineers are more intimately associated with the advancement of bridge design than Isambard Kingdom Brunel. Below is his Royal Albert Bridge (1854–59), constructed of wrought iron using a pair of lenticular trusses supported on masonry piers.

Isambard Kingdom Brunel

No other individual in Britain made such a marked contribution to civil engineering as Isambard Kingdom Brunel (1806–59). His prodigious list of works includes tunnels, bridges, boats, buildings and railways. Brunel gained his early engineering experience working on the construction of the Thames Tunnel for his father, the engineer Marc Isambard Brunel. Brunel's diverse career was dominated by his work on the Great Western Railway between London and Bristol. He also designed transatlantic ships, including the first metal, engine-powered and propeller-driven ship, the SS *Great Britain* (1843).

Chepstow

Brunel's economical design for the 92 m (302 ft) span railway bridge over the River Wye at Chepstow (1852) uses a bowed tubular girder to support a rigid iron deck using iron chains. His novel design was a prototype for the larger and more accomplished Royal Albert Bridge.

Wharncliffe Viaduct

One of Brunel's first designs for bridges along the Great Western Railway was the 270 m (885 ft) long Wharncliffe Viaduct (1836–37) west of London, between Hanwell and Southall. The brick bridge comprises eight semi-elliptical arches supported on hollow tapered piers.

Hungerford Bridge (above)

The Hungerford Bridge (1845) was an iron suspension footbridge across London's River Thames. The bridge was purchased by the South Eastern Railway Company in 1859 who dismantled it when building Charing Cross Station. The new railway bridge used Brunel's brick piers and the bridge's iron chains were reused on Brunel's famous Clifton Suspension Bridge (see pages 198–199).

Royal Albert Bridge (below)

The Royal Albert Bridge (1854–59) is a railway bridge that links the counties of Devon and Cornwall across the River Tamar. The wrought iron structure comprises two lenticular trusses (see page 47), each 138 m (453 ft) long. These are connected to approaches made of a continuous iron plate beam supported on stone piers, giving the bridge a total span of 666 m (2,185 ft).

John A. Roebling

Brooklyn Bridge
The largest of
Roebling's bridges
is the 486 m
(1,594 ft) long
Brooklyn Bridge
(1883). When
completed it
was the longest
suspension bridge
in the world. In
1869 Roebling
died from tetanus
poisoning after a
ferry crushed his
foot while doing
survey work. His
son, Washington
Roebling, took over
the job, but he was
later paralysed after
suffering from
decompression
sickness while
working on the
caissons that kept
the foundations free
of water during
their construction.
John Roebling's
daughter-in-law,
Emily Warren
Roebling, then
continued
the project.

John A. Roebling (1806–69) spent the first 25 years of his
life in Germany, where he studied and trained as an
engineer. In 1831 he left for the United States where, after
a few years working as a farmer, he returned to engineering.
He specialised in suspension bridges, having studied them
in Germany, and manufactured wire cables on his
Pennsylvania farm. His first suspension bridges were
aqueducts, after which he went on to design much larger
bridges, culminating in his most famous work, the
Brooklyn Bridge across New York's East River.

Niagara Falls Suspension Bridge

In 1851 Roebling started work on his first major suspension bridge, which crossed the Niagara River as part of the new railway between New York and Canada. The 251 m (823 ft) span bridge was groundbreaking in its use of metal wire cables to support the double deck, which carried trains above and vehicles below.

Overhead view of Delaware Aqueduct

NW elevation section

New York   *Pennsylvania*

Delaware Aqueduct (above)

Roebling honed his understanding of suspension bridges on four aqueducts he designed for the Delaware and Hudson Canal from 1848. This particular bridge has four separate spans supported on three piers.

Roebling Suspension Bridge (below)

Roebling's 322 m (1,056 ft) long Cincinnati–Covington Bridge (later renamed the John A. Roebling Suspension Bridge) over the Ohio River was the longest suspension bridge in the world when it opened in 1866, and was a precursor to his Brooklyn Bridge.

Robert Maillart

Salginatobel Bridge
Salginatobel Bridge
(1930) shows
Maillart's innovative
use of concrete.
His intuition for
and understanding
of concrete's
behaviour made his
structures not only
beautiful but cheap
and quick to build.

The first person to fully embrace concrete's aesthetic and structural potential in bridge design was the Swiss engineer Robert Maillart (1872–1940). Although Maillart trained at Zurich's Federal Institute of Technology, he was more an intuitive engineer than a methodical one. His ability to manipulate concrete, which at the time was a new and little-understood material, was revolutionary. Schwandbach Bridge and Salginatobel Bridge (see pages 128–129) are his most renowned and radical works, influencing future generations of bridge designers.

Stauffacher Bridge

One of Maillart's earliest bridges was the Stauffacher Bridge (1899) in Zurich, Switzerland. The bridge appears to be constructed out of stone but is in fact fabricated from reinforced concrete using a hollow box above a three-hinged arch. The deception illustrates Maillart's early work before he fully grasped the structural freedom and simplicity of reinforced concrete.

Bohlbach Bridge

In the Bohlbach Bridge (1932) Maillart developed the deck-stiffened arch technique. The reinforced concrete deck sits on a slender arch supported by slim piers the same width as the arch and deck. The road's banked curve, which counters the centrifugal forces created by traffic, is accommodated by varying the width of the arch.

Schwandbach Bridge

The zenith of Maillart's use of the deck-stiffened arch was at Schwandbach (1933), where the polygonal arch beneath the stiffened deck is 200 mm (7⅘ in) thick. As with Bohlbach, the road passing over the bridge is curved, causing the width of the arch to vary from 4 m (13 ft) in the centre of the arch to 6 m (20 ft) at the abutments.

Santiago Calatrava

The work of Santiago Calatrava (b. 1951) seamlessly traverses the disciplines of engineering, architecture and sculpture. His famous structures, which include bridges, theatres, skyscrapers, schools and railway stations, have transformed and reinvigorated public appreciation of engineering in the 20th and 21st centuries. His bridges are celebrated for their distinction and sculptural quality.

Jerusalem Chords

The cantilever spar cable-stayed bridge, which Calatrava first used in his design for Seville's Puente del Alamillo (1992), has become one of his trademarks. The technique is a variant of the standard cable-stayed bridge in which the mast is angled to counter the load of the bridge, requiring fewer cable stays. Calatrava used the cantilever spar method for the Jerusalem Chords Bridge (2005–08) (carrying trams, vehicles and pedestrians) because of the bold visual statement it makes.

Sundial Bridge

In his design for the cycle and footbridge over the Sacramento River at Redding in California, Calatrava transforms the 66 m (217 ft) mast into a sundial's gnomon. The clock face forms a plaza on the north side of the bridge. The 14 cable-stays of the Sundial Bridge (2004) use more than 1 km (³⁄₅ mile) of cable to support the 213 m (700 ft) span.

Ponte della Costituzione

The Ponte della Costituzione (2007–08) is an arched truss comprising five arched members: one in the centre, two underneath the bridge and two either side. With an 80 m (262 ft) span, these members are connected with steel tubes and plates that form a sculptural ribbed structure supporting the stone and glass steps of the bridge. (See also pages 154–155.)

Samuel Beckett Bridge

Calatrava modified the cantilever spar cable-stayed bridge in his design for Dublin's Samuel Beckett Bridge (2009) by using an arced tubular steel mast supported by cables on both sides. The road and pedestrian bridge swings 90 degrees to allow boats on Dublin's River Liffey to pass. Although the bridge was designed to be a modern landmark, its form also evokes the traditional Irish harp.

Gustave Eiffel

Having trained at the prestigious École Centrale des Arts et Manufactures in Paris, Gustave Eiffel (1832–1923) went on to become France's most famous engineer. Although he designed many groundbreaking structures, the most famous of which is the tower he designed and built for the Paris Exposition Universelle in 1889, which bears his name. Intended to be a temporary structure, the Eiffel Tower captured the hearts and minds of Parisians and is the city's most famous icon. Eiffel also designed the armature inside the Statue of Liberty (1886), which was a gift from France to America.

Garabit Viaduct

The Garabit Viaduct (1884) across the Truyère Valley in southern France advances Eiffel's earlier designs, including the Dom Luís I Bridge (1875) in Portugal (see page 64). The 565 m (1,854 ft) long steel bridge is dominated by the 165 m (540 ft) span crescent arch, on which the 124 m (407 ft) high trussed deck sits. It was the highest bridge in the world at the time.

Maria Pia (left)

Another of Eiffel's early designs is the two-hinged crescent arched Maria Pia Bridge (1877) in Portugal. As with the Dom Luís I Bridge, the wrought iron trussed deck of the Maria Pia is incorporated into the crown of the arch, whose 160 m (525 ft) span was then the longest in the world. Eiffel was a pioneer in the use of trusses, which he exploited as they reduced wind resistance compared with solid beams.

Zrenjanin (right)

The Great Bridge (1904) in the city of Zrenjanin in Serbia uses a small steel truss to cross the Begej River. Eiffel designed the bridge so that it could be raised and lowered to allow river traffic to pass underneath. The bridge was replaced in 1969, but plans are underway to rebuild it.

Cubzac-les-Ponts

The Cubzac-les-Ponts (1883) in southern France comprises a series of box trusses supported by seven pairs of cross-braced round wrought iron piers on stone footings. The approaches are built on a masonry viaduct with longitudinal pointed-arch vaulting. Eiffel's road bridge is 550 m (1,800 ft) long and weighs more than 3,000 tonnes.

ENGINEERS

Benjamin Baker

Partnership

After working for Wilson, Baker joined the engineer John Fowler, and they went into partnership in 1875. Together they worked on London's Metropolitan Railway, the world's first urban railway network, and the Forth Rail Bridge (1882–90), Scotland.

As a young engineer, Benjamin Baker (1840–1907) was apprenticed to the renowned Neath Abbey Iron Works in Wales, after which he became a prolific writer and learned teacher of engineering and one of the greatest engineers of the Victorian age. As assistant to Mr W. Wilson from 1860, Baker worked on the construction of London's Victoria Station and Victoria Bridge before working on the first underground railways. His contributions to engineering span the globe and include tunnelling for the London Underground, the Hudson River Tunnel in New York, and the building of Egypt's Aswan Dam. The one design for which he will always be remembered is the Forth Rail Bridge.

Troubled start

The Forth Bridge Company was formed in 1873 and accepted a suspension bridge design by Sir Thomas Bouch, engineer of the Tay Bridge. Unfortunately for Bouch, the Tay Bridge collapsed in 1877, along with public confidence. His Forth suspension bridge was scrapped and Fowler and Baker's design for a steel cantilever and central girder bridge comprising three massive double cantilevers was accepted.

Construction

Fabricated from 65,000 tonnes of steel (the first time that steel was used so predominantly in a bridge), the bridge's double cantilevers sit on granite piers and iron caissons. They are linked by short suspended trusses creating two main spans over 500 m (1,640 ft) long, the longest in the world until 1917.

PART TWO Case Studies

Introduction

A beam bridge is a structure comprising a horizontal element supported at both ends by piers. The weight of objects moving over the beam (live loads) combined with the weight of the beam (dead load) create vertical forces that are transferred through the structure and supported by the piers. The ability of the beam to withstand these forces may be judged intuitively by the span-to-depth ratio, which is defined by the relationship between the depth of its section and the distance between the piers. If the span is too wide and the section too thin, the beam may fail.

There are two types of beam bridge: a simply supported beam and a continuous beam. The former is the basic type, in which a single beam structure is supported at either end by piers. The latter comprises a beam structure that is continuous over the supports. The thickness of a continuous beam can be reduced in section because it works more efficiently by developing hogging moments in its cross section over the piers as well as sagging moments over its span. Beam bridges are among the longest in the world.

Alter ego
The Rio–Niterói Bridge, Brazil, is officially known as the President Costa e Silva Bridge. At more than 13 km (8 miles) long, it was the second longest bridge in the world when completed in 1974.

An Ping Bridge

An Ping Bridge is an old beam bridge built in the 1140s in Fujian Province, southeast China. Extending a distance of 2,255 m (7,400 ft), the 3–4 m (10–13 ft) wide bridge was for many centuries the longest beam bridge in the world, and it was the longest bridge in China until 1905.

Each section of the bridge is made up of six stone beams laid next to one another and supported at either end by a stone pier. Three pavilions and large carved figures once stood along the bridge.

Endurance
Despite its age, the An Ping Bridge is still in use and well maintained by local craftsmen. It is open to pedestrians and cyclists, but not to vehicular traffic.

Beams

The stones that make up the bridge's sections are rectangular beams with a thickness in section large enough to support their own weight of 25 tonnes. The weight of each beam indicates the structure's inefficiency, where much of the structural capacity is used in supporting its own weight rather than the live loads.

Balustrade (right)

A stone balustrade lines the entire length of the bridge. Each deck section has three sections of balustrade. Note how the large load-bearing granite posts sit in the space between the beams on each pier, with two shorter intermediary posts between them.

Piers

The bridge has a total of 331 piers along its entire length. These can be arranged into approximately three different types. Some are rectangular in plan, some are slightly pointed to make them more streamlined, and others, resembling the shape of a boat, are fully streamlined at both ends to reduce resistance to the river. All are built from layers of stone beams perpendicularly arranged to bond the structure together.

Britannia Bridge

Former beam

This image of Britannia Bridge was taken before the bridge was damaged by a fire inside the tubular beams in 1970. It was strengthened using trussed arches, which replaced the tubular beams.

The Britannia Bridge (1850) crosses the Menai Straits between mainland Wales and Anglesey. The original bridge was a revolutionary piece of engineering by Robert Stephenson (1803–59). His radical solution developed a previous design of a road bridge that had used a wrought iron trough. By enclosing the trough, Stephenson was able to create a strong tubular girder through which the railway track could pass. The series of girders were supported on masonry piers, but in 1970 a fire in the bridge weakened the structure and it was rebuilt using trussed arches on the same piers.

Tubular girder

The design and construction of the tubular girder was essential to the bridge's strength. Each section was manufactured from wrought iron sheet riveted together with a smaller boxed section along the top and bottom to provide additional strength and stiffness. The trains ran inside the beam.

Lifting the beams

It was a condition that the bridge construction did not affect the navigation channel beneath. The decision to use beams allowed the piers to be built independently, then each section was lifted into place using hydraulic jacks, a process that took a total of 17 days.

Approach section

70 m (230 ft)   140 m (460 ft)   140 m (460 ft)   70 m (230 ft)

40 m (130 ft)

Approach section

Bridge length

The beams that create each of the bridge's two 140 m (460 ft) main spans were considerably longer than any beam that had been previously built. Each span section weighed over 1,500 tonnes and was supported by tall masonry piers that gave the bridge a clearance of 40 m (130 ft). Two shorter 70 m (230 ft) span sections connect the bridge to the mainland. The total length of the bridge, including the shorter approach sections, is 460 m (1,500 ft).

Tay Rail Bridge

Piers
The arched approaches of the Tay Rail Bridge stand on solid brick piers, while the piers supporting the truss beams are concrete arches standing on two legs that are tied towards the base by a horizontal beam.

The original Tay Bridge, completed in 1878, used a system of simply supported truss beams fabricated from a latticework of cast and wrought iron. The following year the bridge collapsed during a major storm, causing Britain's biggest railway disaster of the Victorian era, in which all 75 passengers and crew were killed. The new bridge (1887) was constructed with two different types of simply supported beam: one was rectangular with the railway track above, and the other arched with the track passing between the arches.

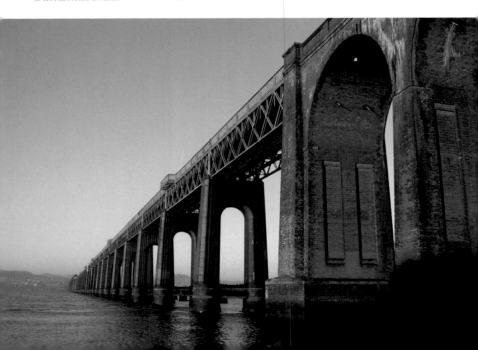

Assorted trusses

Unlike the approaches, which are built from lattice trusses underneath the deck, the central section of the bridge is formed of arched bowstring trusses set above deck and simply supported on piers at either end. This arrangement increases the clearance by extending the distance between piers and the height of the deck above water to accommodate the navigational channel.

Truss beams (below)

On the approaches to the central section of the bridge, the deck is supported on four parallel steel lattice trusses. The trusses are stabilised by cross-bracing and their ends sit on cast iron bulkheads that transfer the load down to two piers.

Joint

Individual members are riveted together to form the latticework. Small vertical members attached to the intersection of the diagonal further strengthen the truss. Expansion joints between the individual beams, where they sit on the piers, indicate that this is a simply supported structure rather than a continuous beam.

Lake Pontchartrain Causeway

The economical benefit of using simply supported beams for very long bridges is nowhere better illustrated than on the Lake Pontchartrain Causeway in Louisiana, USA. The causeway comprises two 38 km (24 mile) long, parallel

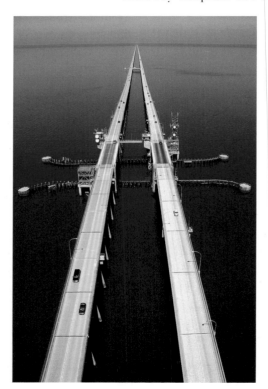

road bridges (northbound and southbound), one built in 1956 and the other in 1969. Except for the central bascule section that allows boats to pass through, the entire length of both bridges is constructed using prestressed reinforced concrete beams simply supported on concrete piers. Each beam was prefabricated and taken to site on a barge from where it was lifted onto the piers by floating cranes.

Prefabrication
The use of prefabricated reinforced concrete sections in the construction of a bridge this long significantly reduces construction times and cost, as each section can be fabricated off-site and then fixed into place swiftly, and with relative ease and economy.

Expansion joints

With so many beams laid end to end it is necessary for there to be a small gap between them to allow the structure to expand and contract. One end of each beam is held securely on a perpendicular cross beam supported on piers with the other end free to expand in the direction of the span.

Expansion joint

Piers (left)

Over 9,500 reinforced concrete piers were used in both bridges. These piers are arranged in pairs, except in the approaches to the central section of the bridge where the taller piers are arranged in threes, increasing the strength of the structure. Each section of the bridge comprises a pair of piers supporting a group of seven beams.

Gateways

Long bridges over wide water channels seldom provide navigational clearance across the full width, since this is not necessary for shipping nor is it economical. Many long bridges over water are therefore low-lying with raised or moving sections that provide a gateway through which large vessels can pass. For this purpose, the centre of Lake Pontchartrain Causeway has a raised bascule section (see page 49).

Chesapeake Bay Bridge

Hybrid structure
The Chesapeake Bay Bridge uses three types of bridge system: reinforced concrete beams (foreground), steel truss cantilever (middle) and suspension (background).

The 28 km (17 mile) crossing of the Chesapeake Bay in Virginia, USA, was one of the most complex bridge systems in the world when completed in 1964. The main section of the bridge is a 975 m (3,200 ft) suspension bridge, with steel cantilevered trusses leading up to it. The system also has two tunnel sections that are 1.5 km (1 mile) long. The majority of the bridge comprises steel truss beams and concrete beams on trestle piers. A second bridge was built parallel to the Chesapeake Bay Bridge Tunnel, which opened in 1999.

Cantilever beam

The approach sections to the suspended portion of the bridge features steel cantilevered trusses. These members act like beams but have a cantilevered form over their support piers, indicated by the profile. The purpose of these members is to raise the deck from the low-lying trestles just 12 m (39 ft) above sea level to the suspended section, which has a clearance of 56 m (184 ft).

Precast beams (below)

Precast reinforced concrete beams create the road deck along the trestle sections of the bridge. Each deck section comprises three beams supported on concrete trestle piers at both ends, which transfer the loads to the ground. Precasting allowed the beams to be manufactured on shore and transported out to sea where they were lifted into place by a floating crane.

Bridge profiles (below)

The four separate sections of bridge are linked by tunnels and artificial islands which, combined with approach roads, make the entire bridge and tunnel system 37 km (23 miles) long. The total length of the bridge from shore to shore is 28.3 km (17⅗ miles).

Bridge tunnel profile

Thimble shoal tunnel

Chesapeake channel tunnel

28.3 km (17⅗ miles) shore to shore

Metlac Bridge

Extended deck
The road deck of the Metlac Bridge extends beyond the edge of the beam and is supported on cantilevered brackets projecting from the side of the beam. The projection of the railway deck is less pronounced.

The railway bridge and road bridge that cross the Metlac River in Mexico demonstrate two different types of continuous beam. The 128 m (420 ft) high steel road bridge was the first to be constructed and opened in 1972. It was built using a continuous steel beam carrying four lanes of traffic supported on a series of concrete piers. The longest span between its two piers is 122 m (400 ft). The 131 m (430 ft) high, two-track railway bridge opened in 1984. The deck is a concrete box section. The longest span between its five piers is 90 m (295 ft).

Beam and pier (left)

The narrow beam of the railway bridge is supported on single piers. The piers of the wider road bridge are arranged in pairs linked at the top by cross beams. The load is transferred directly into each pier through the steel girders that form the sides of the beam. The girder is strengthened at its base by a slight thickening of the bottom section of the I-beam as it meets the pier.

Continuous beam (right)

The rail bridge's deck is a continuous concrete beam. The piers taper towards their base and were built first, then each beam was constructed as balanced cantilevers before being connected in the centre of the spans.

Steel girder (left)

The depth of the steel girder is dictated by the length of the span between the support piers. This continuous steel girder is fabricated from individual sections, which are then welded together and stiffened on the inside by the lateral cross-bracing. Lateral and diagonal cross-bracing stabilises the steel girders.

Introduction

Vertical forces acting on an arch are transferred through its curve into the abutments at either side. The strength of the bridge relies on the shape of the arch and the ability of the abutments to resist the vertical and horizontal thrust forces. Whether pointed, round or segmental, all shapes of conventional arch perform the basic function of transferring vertical forces through a curved or stepped profile to the abutments. Different types of arch include deck-arch bridges, which carry a horizontal deck directly over the arch; through-arch bridges, which support the horizontal deck between the abutments by suspending it from the arch; and half-through arches, which use a similar principle, but in this instance the deck pierces the arch above the abutments. Tied arches do not rely on the abutments at all to resist the horizontal forces in the arch. Instead, these forces are resisted as tensile forces in the deck, which is tied to the arch, in the same way that a bow maintains its shape by being tied at either end to the bowstring. Unlike other types of arch, the tied arch maintains its structure independently of the abutments.

Pont du Gard
Note how the supporting piers in the Pont du Gard, France, are aligned vertically from top to base. To maintain these proportions, the widest of the larger arches has four small arches above it where the others have three.

Alcántara Bridge

Voussoirs
Note how the voussoirs in each of Alcántara Bridge's round arches are arranged radially, contrasting with the horizontal layers of stone used in the rest of the structure.

Inscribed on the central archway of the bridge across the Tagus River in Spain, built under the orders of the Roman Emperor Trajan, is (in Latin) the quotation 'I have built a bridge that will last forever'. Despite some incidents of destruction in recent centuries, Trajan's ambitious claim has proven correct. The stone structure completed in 106 CE and comprising six round arches (three of which span the river), has stood for nearly two thousand years.

Streamlining

The stone piers supporting the arches mid-stream are necessarily substantial. To reduce resistance when the river is in flood the upstream face of each pier is pointed to make it streamlined. Note how the side of the pier facing downstream has not been streamlined. Rounded buttresses extend from each of these pointed profiles to the full height of the bridge to provide lateral stiffness to the sides of the bridge.

Triumphal arch

A small triumphal arch stands in the centre of the bridge. The sides of this smaller lateral arch stand on the buttressing that provides lateral strength to the structure and transfers forces through the piers to the stone foundations.

Symmetry

The six round arches have been built symmetrically. The outermost arches are the smallest, followed by two medium-sized arches and, in the middle of the bridge, the two largest arches, which span the river.

Zhaozhou Bridge

Decoration
Traditionally, Chinese bridges were often adorned with decorative motifs or small pavilions. Here, ornate stone balustrades are topped with carved figures on the posts.

Built between 595–605 CE in China's Hebei Province, the Zhaozhou Bridge is the oldest stone open-spandrel bridge in the world. The 9 m (30 ft) wide and 50 m (165 ft) long bridge is a segmental arch with a span of 37 m (121 ft). The bridge's lightweight design is remarkable for its age. The shallow arc and the open spandrels decrease the bridge's dead load by minimising the stonework, which consequently reduces the size of the abutments and the force on the arch. The combined use of limestone and iron dovetail joints is also very innovative.

Length-to-span ratio (above and left)

The arc of the Zhaozhou Bridge is approximately 87 degrees and has a radius of 27 m (89 ft). The shallow arch gives the bridge a length-to-span ratio of approximately 1:1, making it lightweight when compared with deeper arches where more material is required to counter the steep rise of the deck. The side elevation of the bridge reveals how little stone is needed to create the deck above the slender arch.

Open spandrels (below)

The two pairs of double round-arched open spandrels either side of the arch reduce the weight of the bridge by approximately 700 tonnes, ensuring it does not collapse under its own weight, and also help the flow of flood waters through the structure.

Dovetail joints (above)

Iron dovetail joints between the 28 curved limestone voussoirs reinforce the structure, and prevent the arch segments from separating.

Luguo Bridge

Dry river
Until relatively recently, the Yongding River passed underneath the Luguo Bridge, but water diversion in the Beijing region has meant that the riverbed is now invariably dry.

Originally built in the 12th century, the Luguo Bridge in China was reconstructed in 1698. It is known outside China as the Marco Polo Bridge, because the famed Venetian traveller is said to have praised it highly. The 266 m (873 ft) long and 9 m (30 ft) wide structure comprises 11 segmental granite arches on ten stone piers between abutments at the bank. Carved lions sit on the pillars of the stone balustrade that lines the bridge's deck. It was on this bridge in July 1937 that the 'Incident' between the Chinese and Japanese armies took place, marking the start of the Second Sino–Japanese War.

Segmental arch

A segmental arch has a curve that forms only a portion of a semicircle. The Luguo Bridge has 11 segmental arches constructed of wedge-shaped voussoirs that transfer the load into the piers. The spandrels between the arches are filled with horizontal bands of thin stone.

Voussoir

Streamlined piers

The piers on the upstream side of the bridge have been designed with pointed heads to protect them from the force of the river that once flowed beneath the bridge. Iron bars have also been added to the heads to help protect the stonework from the water and ice.

Ornament

There are so many lions carved into the 280 balustrade posts that the exact quantity is not known, though it is somewhere between 482 and 496. Each lion is different and its posture denotes its age of production.

Pont del Diable

Reconstruction
Built on Roman foundations and having survived seven centuries, the medieval Pont del Diable spanning the Llobregat River was destroyed in 1939 during the Spanish Civil War. It was rebuilt in 1965 in a manner faithful to the previous design.

A bridge has crossed this section of the Llobregat River in Catalonia, Spain, since Roman times. Medieval stone masons used the Roman foundations to build a new bridge in 1283. The bridge is distinctive for having two different types of arch. The smallest is a narrow round-arched opening in the spandrel at the eastern end of the bridge. The medium and largest arches are pointed arches. The bridge's shape reflects the loads on the structure, which are smallest at the centre of the main span where it is thinnest and increase towards the outer spans, which are supported by the abutments.

Chapel (right)

A chapel has been built on the apex of the pointed arch. The round-arched stone structure is a distinguishing feature of the bridge and mirrors the triumphal arch on the eastern abutment. The shallow depth between the arch and the chapel's pitched roof helps to reduce the load over the main arch.

A Roman triumphal arch stands at the eastern end of the bridge. It was common for such features to appear on important civic structures throughout the Roman Empire. Other Roman remains were used in the foundations, on which the medieval bridge was built.

Pointed arch (left)

The main arch has a span of over 37 m (121 ft) and its pointed form reflects the influence of Gothic architecture at the time it was constructed. The smaller pointed arch has a shorter 19 m (62 ft) span.

Ponte Vecchio

Living bridges

The Ponte Vecchio has hosted all kinds of shops, from florists to tanners. In 1593 only jewellers were permitted on the bridge. Today people live in apartments above the shops, which overhang the sides supported by struts.

An advantage of segmental arches or flat arches over round arches is their more efficient span-to-rise ratio. By using segmental arches instead of the round arches favoured previously by the Romans, the 14th-century builder, Taddeo Gaddi, was able to build the Ponte Vecchio on just two piers across the Arno River in Florence, Italy, in 1345. Having fewer piers was more efficient, as there was less obstruction for boats and flood waters. The width of the bridge's deck is 32 m (105 ft), and has always supported shops or market stalls. From the late 16th century to the early 19th century other structures have continued to be added to the bridge.

Central span

Wooden supports

Shops

Stone piers

Span-to-rise ratio

The central span (30 m/98 ft) is wider than the outer spans (27 m/89 ft). The rise of the arches is between 3.5 m (11½ ft) and 4.4 m (14½ ft), giving the arches a span-to-rise ratio of approximately 5:1, reducing the number of piers required to bridge the river. This reduces the bridge's resistance to flood water and minimises obstruction to river navigation.

Profiled pier

The face of each pier is profiled to reduce resistance against the water flow, particularly when the river is in flood. The bridge was nearly destroyed in a flood in 1966: it swept through many of the buildings, but was not high enough to reach the corridor marked by a series of small windows that passes over the top. The three large windows in the centre were created by Benito Mussolini in 1939.

Pulteney Bridge

Picturesque
Pulteney Bridge is one of just four bridges to have buildings along both sides of its entire length. The south façade forms a picturesque view of the bridge with the three-tiered waterfall in the foreground.

Pulteney Bridge (1773) over the River Avon in the centre of Bath, was Britain's answer to Italy's covered Renaissance bridges and, in particular, Andrea Palladio's unrealised design for the Rialto Bridge over Venice's Grand Canal. Inspired by the reinvention of classical architecture, the designer, the Scottish architect, Robert Adam, designed the arched stone bridge to connect the city of Bath with the undeveloped land across the river that existed at the time. The bridge crosses the river on two streamlined piers supporting three round arches.

Shops (above)

The inside façade of the bridge facing the road is lined with a row of small shops. The architecture of these façades has been much altered since the bridge was originally built.

Evolution (above)

Adam's original design of Pulteney Bridge lasted just 19 years. The façade was altered in 1792 to accommodate enlarged shops and in 1799 severe floods damaged the bridge, requiring it to be substantially rebuilt. Subsequent incremental developments have seen the shops extend out from the north façade on cantilevered supports. The result is a disorderly appearance that contrasts with the original symmetrical design that can still be seen on the south face.

Urban planning

The construction of Pulteney Bridge was an important element of Bath's substantial urban development in the 18th century. Much of the new city was designed along classical lines by the father and son architects both named John Wood. Adam continued this tradition in his original design.

Maidenhead Railway Bridge

Engineering marvel
The width of the two main spans of Brunel's bridge was an engineering marvel for its time. Three of the 13 arches of the older Maidenhead road bridge (1777) can be seen in the background.

In his role as chief engineer to Britain's Great Western Railway, Isambard Kingdom Brunel designed the world's longest and flattest arched bridge in Maidenhead, just west of London, in 1838. The Great Western Railway's board were not convinced that such a shallow-arched structure could remain standing let alone withstand the heavy load of the railway. Consequently, they insisted the wooden formwork supporting the two arches during construction be left in place. Brunel is said to have lowered it slightly so that it had no structural function and soon afterwards it was washed away in a flood. The distinctive brick bridge has remained standing ever since.

Arch series
The entire bridge comprises eight arches. The two shallow arches forming the main spans over the river are flanked by three smaller round arches on land.

Arch width (right)
The original design accommodated two tracks of Brunel's wide-gauge railway, which was later abandoned in Britain for the narrower standard gauge. The brick bridge was later widened to carry four tracks of the narrower gauge.

Flat arch
The combined length and flatness of the arch allowed the bridge to cross the River Thames in just two spans of approximately 39 m (128 ft) each while preventing a steep gradient in the railway that would have been created by a taller arch. The total rise measures approximately 7 m (23 ft).

Plougastel Bridge

Reinforced concrete was used to build the three large segmental arches of the Plougastel Bridge (1930) crossing the River Elorn in northwestern France. Designed by the civil engineer, Eugene Freyssinet, the double-decker bridge once carried both vehicular traffic and trains. The trussed section is a lower deck for the railway crossing. The main deck is supported on the tops of the arches. One of the arches was destroyed by the German army in 1944, but it was later rebuilt and widened.

Vital statistics

Each arch has a span of 188 m (617 ft) and rises over 27 m (89 ft), giving them a span-to-rise ratio of 7:1. The total length of the bridge is 888 m (2,913 ft).

Main deck (right)

Five parallel reinforced concrete beams support the bridge's upper deck along the approach section. Note the flange (ridge) between the outside of the external beam and the deck, which transfers the loads from the overhanging 9 m (30 ft) wide deck into the beam.

Flange

Hollow box

The route of the lower deck appears to be blocked by the crown of the arch, but it does in fact pass through the structure, because it is hollow. The hollow box arch was constructed using a floating section of falsework. Once the concrete has acquired an appropriate strength as it cures, the falsework can be removed.

Double decker

The series of trusses beneath the main deck create a secondary deck beneath. Both the main deck concrete structure and the steel trusses beneath span between the slender vertical concrete piers in the open spandrel, which transfers the load into the arch. Note the transverse bracing beneath the railway deck, which stiffens it between piers.

Salginatobel Bridge

Robert Maillart's 133 m (436 ft) long bridge (1930) across the Salgina Valley in Switzerland heralded a revolution in the use of reinforced concrete in bridge construction. The bridge is renowned for the economical way its reinforced concrete deck bears heavy loads over its 90 m (295 ft) span. Concrete hinges were built into the structure at the foundations and at the crown of the arch, creating a three-pinned structure. The bridge has a slight gradient, creating a height difference between both ends of approximately 4 m (13 ft).

Three-pinned arch

Three-pinned arches have several advantages over two-pinned or fixed arches, in which horizontal thrusts and bending moments at the centre are difficult to determine. A three-pinned structure can be calculated accurately. It is also better at accommodating movement, caused either during the construction process or by expansion, contraction or creep during its working life.

Structural efficiency (below)

The individual elements of the arch reveal the structure's efficiency. The arch increases in width at the abutment to dissipate the loads into the foundations. The walls are inset from the arch and deck, and increase in height to the quarter span, reflecting the bending moment distribution. The evenly spaced columns supporting the deck are linked transversely by slender slabs that stiffen the structure.

Construction

The 3–4 m (10–13 ft) wide plate of the concrete arch was constructed in situ on falsework in less than two days, followed by the arch walls, columns and deck. The falsework itself is a notable piece of engineering that cantilevers out from the steep valley walls.

Sydney Harbour Bridge

Arch thickness
The thickness of the Sydney Harbour Bridge arches varies from 18 m (59 ft) at the crown, where the loads are least, to 57 m (187 ft) at the approach road end, where the loads are greatest.

Standing 134 m (440 ft) above sea level and with a span of 503 m (1,650 ft), Sydney Harbour Bridge (1932) is still among the world's tallest and widest steel arch bridges. Fabricated from 53,000 tonnes of steel and more than six million rivets, the two-pinned arch structure is formed of two trussed arches joined by crossed-bracing. The deck is suspended from the arch on hangers that descend from beneath the vertical elements of each truss section. The width of the deck is substantially wider than the arch structure and carries eight lanes of traffic, two railway tracks, a cycle lane and a footpath.

Through arch

The deck of the bridge intersects the arch above the abutments, making it a through-arch bridge. Note how the granite-clad concrete pylons are detached from the arch structure except at their base and therefore serve no specific structural function.

Half arch construction

Each side of the arch was built up from the abutments using electrically powered cranes that crawled up the arch as it grew. To prevent the half arches from toppling during construction, they were held back by 128 steel cables anchored into the ground. The arch is an example of the Pratt truss arrangement, where the diagonal members in each of the 28 truss sections slope down towards the centre of the arch.

Abutments

The foundations were dug to a depth of 12.2 m (40 ft) and excavated 122,000 m³ (4.31 million cu ft) of rock. They were filled with high-grade concrete to support the four massive pins on which the entire structure sits. Each pin is 4.2 m (13⅘ ft) wide and 36 cm (14 in) thick, resisting a thrust of approximately 20,000 tonnes. The pins are also hinges that allow the bridge to expand, contract and rotate, which can alter the height of the bridge by up to 18 cm (7 in).

New River Gorge Bridge

Weathering

The New River Gorge Bridge was built using COR-TEN steel, a type of weathering steel that does not require painting. The result is a natural rusty coloured finish in keeping with the natural environment in which the bridge is located.

When completed in 1977, the New River Gorge Bridge in West Virginia, USA, was the longest single-span arch bridge in the world and the highest arch over water. The two-pinned, steel truss segmental arch has a span of 517 m (1,696 ft) and rises to 267 m (876 ft). The arch supports a continuous truss deck with two expansion joints over the piers rising from the abutments. The steel piers supporting the deck are evenly spaced and rise from the valley sides and across the arch. The arch was constructed simultaneously from both sides, each side being anchored by steel wires until the arch met in the middle.

Arch section lengths (above)

The profile of the main arch is slightly tapered, being thicker at the base, where the loads are greatest, and thinner at the crown, where the loads are least. The piers are similarly tapered from the base to the top. The cross-braced supporting columns are connected to the arch at every third truss section. Note how these arch sections vary in length to ensure a uniform distance between the piers is maintained along the length of the bridge.

Foundations (above)

The foot of each arch comprises a steel-pinned connection to allow for rotation generated by loading and expansion and contraction of the steel. Note how the structure has been strengthened where it meets the pin joint by building up a series of riveted steel plates.

Deck truss

The road deck is constructed on a continuous through-type subdivided Warren-truss beam supported on the tapered columns at every third truss section. The deck is stiffened by lateral and longitudinal cross-bracing underneath.

Bloukrans Bridge

Environmental impact

An arched bridge was selected to cross the Bloukrans River so as to minimise the impact on the surrounding natural environment. The rocky valley sides were ideal for supporting the thrust forces in the arch.

Towering 216 m (709 ft) over the Bloukrans River in South Africa, the Bloukrans Bridge (1984) is the highest single-span arched bridge in Africa. The 272 m (892 ft) span reinforced-concrete deck-arch structure stands on foundations built into the valley walls at a height of 150 m (490 ft) above the valley floor. Pairs of slender columns support the 450 m (1,480 ft) long and 16 m (52 ft) wide concrete deck. The arch has no hinges at its feet or crown, making it a fixed arch. The space between the crown of the arch and the deck accommodates the world's highest bungee jump (216 m/709 ft).

Cantilever construction (above)
The reinforced concrete bridge
was constructed using the suspended
cantilever method. Each arch section was
constructed incrementally from either side
of the valley and anchored to the bedrock
with steel cables. Once the arch met in the
middle, the restraining cables
were removed. The supporting
columns and deck are then
constructed on top of the arch.

Open spandrels (below)
The 46 columns that support the deck
are arranged in equally spaced pairs
19 m (62 ft) apart. There are three
different sizes of column, though all
are approximately 2.5 m (8 ft) wide
and 1 m (3 ft) thick. The almost
imperceptible differences are testament
to the quest for efficient design.

Bridge cross section

Deck

Columns

Main arch
cross section

Section (left)
This section through the bridge reveals
the hollow box structure of the 12 m
(39 ft) wide main arch supporting the
pair of columns. Each 2.5 m (8 ft) wide
column supports an 8 m (26 ft) wide,
hollow deck section.

Chaotianmen Bridge

Chords

The principal members of a truss are the chords, between which the vertical and diagonal members are arranged. Here, the upper and lower chords are red.

In 2009 the Chaotianmen Bridge across the Yangtze River in the city of Chongqing (the world's most populous city) in southwest China became the world's longest arch bridge. The main span of the half-through arch is 552 m (1,811 ft) long and 142 m (466 ft) high and carries a double deck. The bridge's main span is a two-pin, steel-trussed arch supporting a pair of steel decks that are stiffened with lateral transverse bracing.

Truss arch

The bridge's wider upper deck extends beyond the sides of the arch. The continuous steel-truss structure can be seen passing through the deck. Note how the truss's diagonal members change direction where the arch meets the ground. This change of direction maintains tension in the diagonal members that carry the loads back to the support. The top and bottom members of the truss are fabricated from welded box steel sections.

Double deck (below)

The 36.5 m (120 ft) wide top deck carries six lanes of road traffic and two pedestrian paths on each side. The 29 m (95 ft) wide lower deck carries two railway tracks and four lanes of traffic. The upper deck is stiffened with longitudinal 'U'-shaped ribs and the lower deck, which is suspended from the upper deck with tapered steel columns, is braced on its underside with transverse and diagonal steel members.

Deck cross section

Pedestrian walkways

Two alternative traffic lanes

Two alternative traffic lanes

Two lanes of light rail

Upper deck carrying six lanes of traffic

Foundations (left)

Note how the foundations are required to take only a vertical load because the truss is tied. This is in contrast to abutment-type foundations for arch bridges, which need to resist horizontal thrust.

Introduction

Truss bridges rely on the inherent strength of the triangle, whose individual members are subject to compression, tension, or both (though not at the same time). The two principal elements of a truss are the two chords (upper and lower), the outside members of a truss, which extend from one end of the truss to the other and are connected by a series of smaller vertical or diagonal members in compression or tension. There are many configurations of truss, which create a wide range of different patterns and appearances on bridges, from simple triangular types to very complex latticework. The amalgamation of truss elements into an entire structural framework performs the function of a beam, arch or cantilever, and provides support for the bridge.

Truss bridges can therefore be as simple as a triangle carrying a footbridge over a stream or as complex as a series of cantilevered sections carrying a multi-deck road-and-rail bridge across the sea.

Kingston–Rhinecliff
The Kingston–Rhinecliff Bridge (1957) in New York, USA, comprises a series of different truss types that create ten spans and carry two lanes of traffic across the Hudson River.

Pont-y-Cafnau

Bridge of troughs
A column rose from the structurally rigid point of the Pont-y-Cafnau where the apex of the truss met the parapet to support an overhead trough that once carried water and limestone to the ironworks.

Built in 1793, the Pont-y-Cafnau over the River Taff in Wales is an example of an early type of truss bridge and is the world's oldest surviving iron railway bridge. Built to serve a local ironworks, the bridge once comprised two independent decks: a railway at ground level and an aqueduct above. The 14 m (46 ft) wide deck is a cast iron hollow box, 2 m (6½ ft) wide and 0.6 m (2 ft) deep, supported by three lateral beams connected to two large A-frame trusses on either side. The apex of each truss meets the top of the parapet in the middle of the span.

A-frame truss (right)

The horizontal tension member of the A-frame ties the two diagonal sections together at the same height as the bottom of the deck, with the lateral beams supporting the deck socketing into its sides. The vertical member in the centre of the A-frame extends above the parapet to provide a bracket in which the wooden columns supporting the aqueduct were fixed.

Horizontal member

Vertical member

Diagonal member

Carpentry in iron (below)

The bridge's designer, Watkin George, was a trained carpenter and designed the bridge as if it were to be fabricated in wood. The mortise-and-tenon and dovetail joints in the wrought iron members reflect his former vocation.

Abutments (above)

The valley walls were reinforced with vertical stone abutments. These support the vertical forces from the deck, which sits on top of the abutments, as well as lateral forces in the diagonal members of the A-frame lower down the abutments.

Busseau sur Creuse

Continuous beam
The railway deck of the Busseau sur Creuse Viaduct is a continuous lattice truss beam. This is revealed by the absence of joints along the truss over the piers. A continuous truss minimises the beam's depth and is more efficient than simply supported spans. The beam is supported by five piers and stone abutments at either end, one of which contains three round arches. Four of the six spans are 50 m (165 ft) long, one is 45 m (147 ft) and another is 41 m (134 ft).

Lattice trusses were first patented in the early 19th century. The deck of the railway viaduct in Busseau sur Creuse (1863), France, is laid on top of the beam, whose sides are constructed of closely spaced parallel rows of flat wrought iron members riveted together at 45-degree angles. The piers stand on masonry foundations on the valley floor. The entire bridge is 338 m (1,109 ft) long and weighs over 2,000 tonnes, including 15 tonnes of paint.

Deck, pier and railing

The 5 m (16 ft) deep lattice trusses not only form the walls of the deck, but are replicated across the 8 m (26 ft) width of the deck. Two vertical members strengthen the truss in the areas of high shear force where it is supported on the piers.

Tapering piers

The tapered wrought iron piers decrease in cross section as they rise from the foundations to the deck. The taper is more extreme in the direction perpendicular to the span. The piers are constructed in sections, each one of which is defined by horizontal bands strengthened by diagonal bracing. Note the bolted flange details between sections. The masonry foundation supporting the tallest pier is 18 m (59 ft) high and the height of the bridge at that point is 56 m (184 ft). The wrought iron piers are flanged at their base to provide bolt housings, where they are connected to the masonry piers.

Howrah Bridge

Clearance
The Howrah Bridge's 30 m (100 ft) wide deck (with a 22 m/72 ft central section for traffic between two 4 m/14½ ft pedestrian lanes) has a clearance above water of 9 m (30 ft).

Truss bridges commonly support the deck on, within or at the base of the truss, but the deck of the Howrah Bridge (1943) over the Hooghly River in Kolkata (Calcutta), India, is suspended beneath the large cantilever truss. Made from riveted steel members, the 20,000 tonne structure comprises three main sections: two cantilevers and a suspended central section. Joints are located between these three sections to allow for thermal expansion and contraction. The main span is 457 m (1,500 ft) and the towers of each cantilever section are 83 m (272 ft) tall.

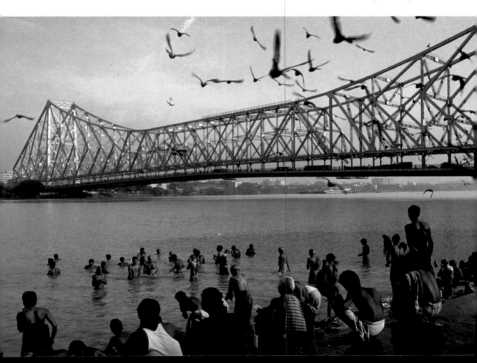

Deck (below)

The deck is suspended from the lower chord of the main structure by 39 pairs of vertical steel trussed members. Compared with the main cantilever members, these 'hangers' can be relatively slender as they are working in tension and not compression, so are not prone to buckling.

Anchor arm (below)

Because the deck is suspended beneath the truss, it reaches land at the tower, leaving the portion of the cantilever on the land-side of the tower free from carrying deck loads. Note the K-bracing diagonals are oriented in the opposite direction to the Minato Bridge (see page 151), therefore the bottom members are acting in compression and the top ones in tension. The bottom diagonals are further braced to withstand buckling forces.

Caisson (above)

The Howrah Bridge has the largest land-based caisson in the world. Caissons are chambers lowered into the river and filled with air to create a space in which workers dig out the soil. The caisson sinks as the soil is removed until they reach a suitable substrate. The bridge's massive caisson was dug to a depth of over 29 m (95 ft). Five hundred people were employed to maintain the pressure inside the caisson to prevent it from filling with water.

Kingston–Rhinecliff Bridge

Cofferdams

Construction on the piers began in 1954. The foundations were constructed using cofferdams, in which large sheets of metal are placed into the riverbed to form an enclosed space. The water is pumped out and the concrete poured in to form the base of the piers.

Originally intended to be a suspension bridge, the bedrock on the banks of the Hudson River was found to be inadequate for anchoring the cables so a bridge comprising a series of deck trusses was selected instead. Opened in 1957, the 10 m (33 ft) wide two-lane steel truss bridge has ten spans: four central spans of approximately 150 m (490 ft); two main spans of 244 m (800 ft); and four smaller spans partially on land. The entire length of the bridge is 2.3 km (1⅖ miles).

Bearing joints (right)

Each of the nine piers between the main abutments is constructed of pairs of cross-braced, reinforced tapered concrete columns. Two bearing joints on top of each pier support the trusses at its outer wall. Eight of the piers stand on deep foundations in the river and one is on land.

Navigation channel (above)

The arched profile of the main navigation channels reveals their longer spans. The truss is deepest at its ends, where the shear forces are greatest, and the profile causes the truss to act as a cantilever over these supports. The deck is supported on top of the truss along the entire length of the bridge.

Subdivided Warren truss (below)

Each section of the bridge is constructed using a subdivided Warren truss. Some sections are simple truss beams and others are cantilever trusses. The joint between trussed sections above some of the piers reveals that the bridge is not a continuous truss over its full length and allows for movement between the bridge sections.

Astoria–Megler Bridge

Engineered safety
The Astoria–Megler Bridge is exposed to harsh natural conditions. This truss bridge is designed to withstand violent ocean storms, and its steel piers stand on elongated concrete footings to protect them from debris carried downstream by flood waters.

The main section of the 6.5 km (4 mile) long Astoria–Megler Bridge (1966) in Washington, USA, is the longest continuous truss bridge in North America. The main truss, which has a different height at either end, creates a navigation channel in the Columbia River with a maximum clearance of 60 m (197 ft) and a span of 376 m (1,234 ft). The approach spans are beam trusses of a subdivided Warren type. Most of the bridge is constructed using low-lying, simply supported, prestressed concrete beams.

Camel-back truss (above)

The entire bridge system comprises a series of different truss structures. On the Washington shore side the bridge is made up of seven simply supported subdivided Warren-type camel-back through trusses. The name camel-back comes from the distinctive hump-backed profile of the polygonal upper chord.

Ramps (below)

One approach to the continuous truss section is a continuous beam truss supported on five concrete piers. The steep gradient of the approach ramp from the high continuous truss down to the low section fabricated from prestressed concrete beams requires some horizontal restraint to secure the deck on support piers.

Tapered piers (right)

K-braced tapered piers provide both lateral and longitudinal support to the continuous truss section. The sharp angle of the taper provides a wide supporting base on a concrete foundation that rises to a knife-edge apex supporting the truss's load and stabilising it against any lateral forces such as wind loads. The knife-edge support enables the truss to rotate about this point in response to loads and expansion/contraction movements.

K-bracing

Knife-edge apex

Wide supporting base

Concrete base

Minato Bridge

Through truss
The 22 m (72 ft) wide double deck is housed inside the truss beneath and parallel to the upper chord, making the Minato Bridge a through-truss bridge. The 36,000 tonne structure has the third longest span of any cantilever bridge in the world.

The Minato Bridge (1974) is a double-deck, steel cantilever through-truss bridge in Osaka, Japan. The two large cantilever sections are anchored on piers on the shoreline and extend over the water to create a span 510 m (1,673 ft) long and 51 m (167 ft) high. The structure of the cantilever sections are K-trusses, evolving into a simple Pratt truss in the suspended truss section in the centre of the span. A cantilever truss was considered the most appropriate configuration for this particular bridge because the poor quality of the subsoil was unable to support the weight of an arched bridge.

K-truss

The K-truss sections are clearly visible either side of the central part of the cantilever. Note how the 'K' is oriented in the direction of the cantilever. This means that the lower leg of the 'K' always works in tension and the upper works in compression. Note also how the central part of the 'K' remains at a consistent height in line with the bridge's lower deck and therefore decreases in height as the truss section narrows, turning into a Pratt truss at the bridge's narrowest central portion. (See also Howrah Bridge, page 145.)

Lateral ties (above)
Lateral Warren-type trusses tie both sides of the bridge together, strengthening the structure and preventing twisting.

Not an arch (above)
Viewed in profile, the two cantilevered sections of the Minato Bridge reveal the length of span that can be achieved compared with an arch, which would have required a much larger structure. The entire length of the bridge, including trussed beam approaches, is 983 m (3,225 ft).

Ikitsuki Bridge

Renewable energy
The velocity of the water current is greater around the Ikitsuki Bridge piers, making them ideal sites for generating renewable energy.

With a main span of 400 m (1,312 ft), the Ikitsuki Bridge (1991) is the longest continuous truss bridge in the world. Truss bridges with longer spans than the Ikitsuki Bridge are not constructed from a continuous truss, but have cantilevered sections supporting a suspended mid-section. The truss sections are a combination of K-trusses and a Pratt truss in the mid-section.

Through truss (right)

The road approaches the bridge at one end over a short section of concrete beams simply supported on concrete piers, before passing through the truss structure, making it a through truss.

Piers (above)

Two concrete piers in the open water between Ikitsuki and Hirado Island in Japan support the bridge at the point where the truss profile rises. Each pier comprises two supporting legs joined by a thin wall of concrete and connected at their top by a lateral load-bearing beam. The corbelled profile of the beam reflects the way in which it cantilevers beyond the piers to restrain laterally the outer edges of the bridge deck.

Continuous truss (below)

A continuous truss is defined as a spanning beam structure that develops hogging moments over the supports and sagging moments in the centre span. Although it appears similar to a cantilever bridge and might have been constructed as two cantilevered sections temporarily, it does not behave like a cantilever. The principal difference is that a continuous truss bridge relies on the structural performance of the central span as part of an integrated structural form.

Ponte della Costituzione

Designed by Santiago Calatrava and opened in 2008, the Ponte della Costituzione over Venice's Grand Canal is a steel, arched truss fabricated from five arched sections: a central arch, two side arches, and two lower arches. The five tubular, steel arches are connected to one another with perpendicular tubular and plate-steel members, creating a complex skeleton structure of ribs along five spines. The central and lower arches form a main spine, which is the principal supporting element, while the two pairs of outer arches provide additional support to the splayed deck.

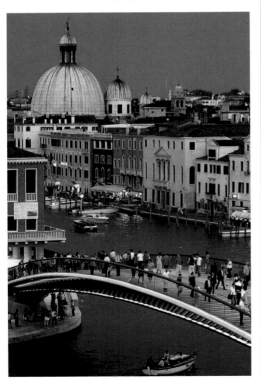

Quiet revolution (left)

Before this bridge was built, none of Venice's bridges exceeded a span-to-rise ratio of 7:1. With a span of 20 m (66 ft) and a rise of just 4.8 m (15½ ft) creating a span-to-rise ratio of 16:1, the Ponte della Costituzione has instigated a quiet revolution in the fundamentals of bridge design in Venice.

Tubular truss

The smaller transverse members, or ribs, are angled radially, pointing down away from the centre of the bridge's span. The absence of any longitudinal triangulation in the truss (that is, a combination of verticals and diagonals) makes this a Vierendeel truss.

Clearance (below)

The single span arch has a 180 m (590 ft) radius creating a shallow arc with a span of 80 m (262 ft) and 7 m (23 ft) clearance. Making the arch stiff enough to work in tension, and not just compression, has resolved two competing needs: pedestrians passing over avoid a steep climb, and boats passing under are not obstructed.

Foundation

The abutments are constructed from reinforced concrete, which is clad in stone. The identical sculpted forms have been carefully designed to complement the gentle arc of the bridge as it touches down on either side of the canal.

Glass deck

The gently rising deck, comprising glass steps with lights within each step, increases in width from the abutments to the mid-span. A parapet of transparent plate glass topped with a bronze handrail maintains the bridge's lightweight and slender appearance.

OPENING & MOVING

Introduction

Moving bridges provide a means of crossing that can be temporarily removed. The most common function of a moving bridge is to allow tall boats to pass, but rarer examples include defensive functions such as drawbridges, transportive functions to move goods across a river, or the protective function of removing the structure from the path of a flooded river. Opening bridges are not confined to any particular structural type, such as arch or beam, and come in all shapes and sizes. More than any other category of bridge, moving bridges present a unique opportunity for design innovation, making them among the most flamboyant bridge type. Moving bridges often also present complex engineering challenges, as their movable elements exert shifting and reversible forces on the structure. Among the many types of moving bridges around the world, there are those that ascend and those that descend, while others swivel, and there are even some that curl or tilt.

Gateshead Millennium Bridge
Completed in 2001, the multi-award-winning Gateshead Millennium Bridge across the River Tyne is the world's first and only tilting bridge.

Barton Swing Aqueduct

Cross-bracing

The cross-bracing in the central section of the truss over the pivot of the Barton Swing Aqueduct provides additional strength over the support. The aqueduct swings at its mid-section on a specially constructed island in the middle of the canal.

The Bridgewater Canal in northwest England was constructed in the mid 17th century to transport coal to Manchester. As one of the earliest transport arteries built in response to Britain's rapid industrialisation it had to successfully traverse existing natural waterways and established transport routes. A stone aqueduct was built for the purpose of crossing the River Irwell, which was replaced in the late 19th century by a swing aqueduct – the first and only one of its kind in the world when it opened in 1894. The replacement was necessitated by the construction of the Manchester Ship Canal in the 1890s, along which much larger ships travelled up the waterway beneath the original aqueduct.

Swing pivot (above)

The 1,450 tonne movable section of the aqueduct is a 71 m (233 ft) long and 7 m (23 ft) wide steel through truss. Note that the truss arrangement looks like a typical simply supported Howe truss. However, the diagonals are in tension not compression because of the cantilever form.

Twin pivot

The aqueduct is not the only swing bridge at this point of the Manchester Ship Canal. An adjacent swing bridge carrying a road over the canal shares the artificial island. The bridges are designed to work concurrently so that in their open positions they are aligned on the island, separated only by the brick control tower.

Tower Bridge

Why the need?
Tower Bridge was designed to allow tall ships to reach the Pool of London between Tower Bridge and London Bridge. Clearance when the bascules are closed is 8 m (26 ft) and 42 m (138 ft) when open.

One of the most famous moving bridges in the world crosses the Thames just outside the eastern boundary of the old City. Tower Bridge, so named for its proximity to the Tower of London on the northern bank, is a combination of a bascule and a suspension bridge and was opened in 1894. The mechanisms for the bascules are housed at the base of the towers, which stand on piers made with 70,000 tonnes of concrete. Following its completion, it was the biggest and most sophisticated bascule bridge in the world.

Bascule and suspension

The 244 m (800 ft) long bridge can be divided into two structural forms: the 61 m (200 ft) central span with its pair of bascules; and two suspended outer sections. The outer sections comprise a deck that is supported by rods hung from suspension trusses. The horizontal tension forces from the suspended section are carried through a pair of lattice trussed box girder beams between the main towers. These beams also house walkways.

Steel truss (left)

Although Tower Bridge has been clad in Portland stone and granite, most of the structural elements are steel and concealed inside. Over 11,000 tonnes of steel were used in the steel frame of the towers, the lattice trusses across the top, and the suspension rods.

Bascule (right)

Each bascule weighs more than 1,000 tonnes. To make the operation of raising them easier and to relieve the stress on the pivot, each bascule is counterbalanced. The counterbalance and the modern hydraulic mechanism used to open the bridge are housed within the piers.

Counterbalance

Middlesbrough Transporter Bridge

Transporter bridges allow low-level navigation across a river while maintaining high-level clearance. This dual function is achieved by having a movable car, or 'gondola', suspended on steel cables from a raised beam. The car travels back and forth across the river using cables and pulleys. The transience and portability of the bridge deck allows the easy and regular passage of large boats under the framing structure while maintaining river crossing. The bridge is suited to sections of river where priority is given to waterborne traffic.

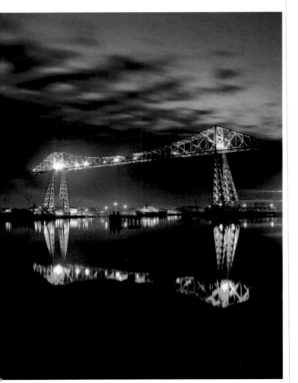

Middlesbrough Transporter
The Middlesbrough Transporter Bridge opened in 1911 and is the last bridge across the River Tees before it reaches the sea. A transporter bridge was chosen because it did not restrict river navigation. Only 20 transporter bridges have ever been built, and only 11 of these survive.

Transporter car

The Middlesbrough Transporter can carry 200 people or nine cars across the river in 90 seconds.

Cantilever (below)

Transporter bridges rely on a high-level structure supporting the cable mechanism from which the car is suspended. Many different methods can be used to achieve this, but the Middlesbrough Transporter uses two 49 m (161 ft) tall steel cantilevered sections supported on tapered pylons that provide a central span of 180 m (590 ft). The cantilever arms are longest across the river and the shorter anchor arms over land are tied to the ground with cables to resist the uplift forces.

Trussed piers (above)

The steel trussed piers supporting the high-level cantilevered sections are tapered asymmetrically so that the inner walls are vertical to allow unrestricted passage of the cables and car, which is an important design consideration.

Anchor arm

Cantilever arms

Anchor arm

Michigan Avenue Bridge

Renamed

The Michigan Avenue Bridge was renamed the DuSable Bridge in 2010, after Jean Baptiste Pointe DuSable, the first non-native settler in Chicago.

Designed to link the northern and southern parts of Chicago across the Chicago River as part of an ambitious early 20th-century urban plan for the city, the Michigan Avenue Bridge (1920) was the first double-decker bascule bridge ever built. Faster non-commercial traffic used the top deck, and slower commercial traffic used the bottom. Pedestrian walkways flank both decks. The bridge comprises two equal bascules, or leaves. Four tender houses stand on each corner of the stone abutments.

Deck truss

The structural form of the double deck in its lowered position resembles a Pratt truss. It functions as both a through truss (with the bottom deck) and a deck truss (with the top deck). Note that the trusses still function in a similar way to a simply supported Pratt truss, with the diagonals in tension, because the cantilever supports are at the ends and not the middle like the Barton Swing Aqueduct (see pages 158–159).

Twin bridges (right)

Each leaf is divided longitudinally so that either side of the bridge can operate independently of the other in the event of one side being struck and damaged by a boat. Each leaf comprises 12 longitudinal steel beams strengthened by diagonal and horizontal cross-bracing.

Trunion bascule

The pivot of each 3,400 tonne leaf is beneath the lower deck and counterweighted by four 1,500 tonne weights that sink into 12 m (39 ft) deep pits as the bridge opens. The bridge pivots on cylindrical pin joints, or trunions, allowing it to move up and down.

Corinth Canal Bridge

Fishing bridge
As the bridge is raised, fish get caught on the deck in the space between the balustrades. The trapped fish are often collected by local children.

The submersible bridge is a comparatively rare bridge type. The method is similar to that of a vertical-lift bridge but rather than the deck being lifted vertically, the deck of a submersible bridge descends to give sufficient draught for vessels. Both ends of the 6 km (3¾ mile) waterway that forms the Corinth Canal between the Corinth and Saronic Gulfs in Greece have submersible bridges (1988) that descend to the bottom of the canal to allow boats to pass.

Bridge up

When fully raised the bridge stands nearly 2 m (6½ ft) above sea level with only a narrow clearance, thereby forming an effective barrier to the canal, and preventing any boats from entering.

Bridge submersing

The main deck comprises four 1 m (3 ft) deep steel girders supporting two lanes of traffic. Treatment of the steel is a critical design consideration to protect it from the aggressive marine environment. Both ends of the bridges are fixed into reinforced concrete piers that house the rising mechanism.

Fully submersed

The bridge maintains a horizontal position as it descends to the bottom of the canal, offering boats the maximum possible clearance beneath their keel.

Erasmus Bridge

Multiple piers
A number of different concrete piers support the 808 m (2,651 ft) long steel deck of the Erasmus Bridge, which carries vehicular traffic, and urban trams, cyclists and pedestrians.

A moving section forms a small portion of the prominent Erasmus Bridge over New Meuse River in Rotterdam, The Netherlands. The 89 m (292 ft) bascule is dwarfed by the 280 m (920 ft) main span suspended by cables from a 139 m (456 ft) high pylon, but it is a vital part of the bridge since it is the only section through which large boats can pass. Opened in 1996, the Erasmus Bridge is the tallest bridge in The Netherlands and its bascule section is the largest and heaviest bascule bridge in Western Europe.

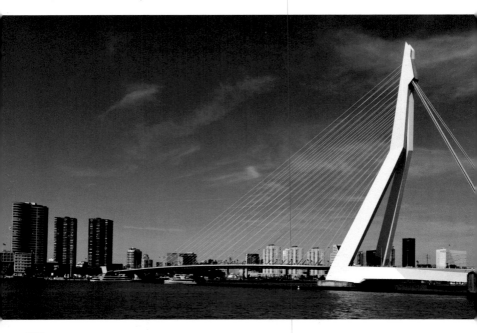

Bascule (below)

The structure of the bridge's large bascule is tapered so that it is thicker nearer the pivot and thinner towards the outer edge. The stronger cross section close to the support, or pivot point, reflects the requirements of the cantilever from. The bascule comprises two principal cantilever arms connected by transverse beams over which the deck is laid.

Tower cable (above)

Nearly 7,000 tonnes of steel were used in the construction of the Erasmus Bridge. Much of this was used in the angled pointed pylon that supports the 32 cable-stays, the longest of which are nearly 300 m (980 ft).

Pylon (below)

The angle of the pylon connected to the horizontal base section of the bridge resists the loads applied by the deck and is tied to the rear of the bridge by eight anchoring stays. Note how it leans against the tensile deck forces.

Gateshead
Millennium Bridge

Award winner
As the world's
first and only
tilting bridge,
the Gateshead
Millennium Bridge
has won many
international
awards for
architecture.

Crossing the River Tyne, and opened to the public in 2001, the Gateshead Millennium Bridge was the world's first tilting bridge. Designed for pedestrians and cyclists crossing between Newcastle and Gateshead, the bridge comprises two counterbalanced steel arches with a span of 105 m (345 ft). One of the arches forms the deck and is connected by steel cables to a supporting arch that acts as a counterweight to assist the tilting action.

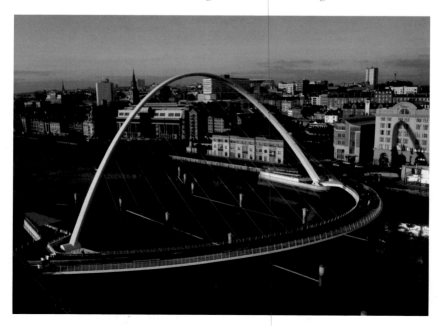

Pivots (below)

Eight electrical motors drive the hydraulic mechanism that tilts the bridge over 40 degrees in less than five minutes. The bridge is also designed to be self-cleaning. Any litter left on the bridge when it is tilting, rolls into traps at the end of the arch.

Deck cables (above)

The deck is divided into two sections. The cycle lane is on the outer curve and is about 30 cm (12 in) lower than the pedestrian deck on the inner curve. The steel cables connect to the inner edge of the deck arch so as not to obstruct head height clearance for pedestrians and cyclists. This means that the deck is cantilevering out from these lines of support.

Tilt

To allow boats to pass, the 800 tonne steel structure tilts from pivot points on abutments constructed from 19,000 tonnes of concrete that project from each bank. As the supporting arch lowers, the deck arch rises until both are equal in height.

The Rolling Bridge

The 12 m (39 ft) long pedestrian Rolling Bridge (2004) over an inlet of the Grand Union Canal in London, comprises eight hinged deck sections with polygonal walls that curl up to form a perfect octagonal structure on one bank. In its unfolded form the bridge is a truss – the handrails forming the upper chord and the deck the lower chord. The slight downward angle of the handrails transfers the loads in compression into the vertical cylindrical posts. Folding takes three minutes and can be stopped at any point during the process.

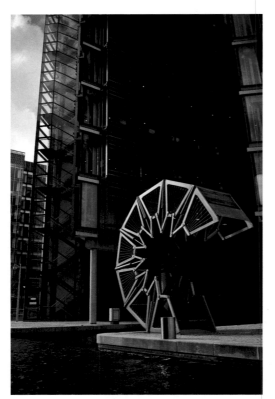

Operation

At the touch of a button, the bridge begins to curl as it rises to a height of almost 10 m (33 ft) before coiling into an octagonal form on one side of the inlet in under three minutes.

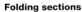

Curling mechanism (above)

The bridge curls by hydraulic actuators
housed in each vertical post driving a metal
cylinder up and out of its casing and forcing
the handrails to fold, during which they
go from being elements under compression
to elements under tension. The vertical
cylindrical posts also change from being in
compression when the bridge is unfolded to
being in tension at certain stages of folding.

Folding sections

The principal elements of the bridge are the
polygonal balustrade panels, the vertical
cylindrical posts, the hinged 'U'-sectioned
handrails (the sides of the 'U' providing
lateral support to the vertical posts) and
the deck sections. To keep the weight of
the bridge to a minimum, the voids in the
balustrade panels are infilled with wires.

Curling sequence

As the bridge starts to curl it transforms
from a simply supported truss to
a cantilevered truss. In the initial phase
of curling, as the handrails pass through
horizontal, the structure arches. A hinge in
the fixed deck section allows this movement.
The bridge rises to its full height, then coils
into an octagonal structure.

Gustave-Flaubert Bridge

A vertical-lift bridge comprises two pylons that support both ends of the deck and raise it in a vertical motion while keeping the deck horizontal. One of the largest vertical-lift bridges in Europe is the Gustave-Flaubert Bridge (2008), the sixth bridge to be built across the River Seine in Rouen, France. The total length of the bridge including approach roads is 670 m (2,200 ft).

Rising system

The 116 m (381 ft) central span of the Gustave-Flaubert Bridge carries twin road decks fabricated from steel girders. These are raised and lowered on a series of steel cables strung over 450 tonne pulley mechanisms, which are housed at the top of each pylon.

Steel rib

Longitudinal beams

Road deck

Each 116 m (381 ft) long steel road deck weighs 1,200 tonnes. The main structure comprises two longitudinal beams that are connected and strengthened by a series of steel ribs, which cantilever from the sides to give a suitable deck width.

Pylons

The pair of 86 m (282 ft) high reinforced concrete pylons each comprises two independent masts joined at the top by the pulley mechanism. This gives structural depth across the pylons by ensuring that they work together. The road decks either side of the pylons are suspended at both ends by two pairs of cables – one pair fixed to the outside of the deck and the other fixed to the inside of the deck.

Pulley system (above)

The butterfly pulley mechanism at the crown of each pylon houses eight winches. Note how the structure of the mechanism is strengthened by being cross-braced. More than 6 km (3¾ miles) of steel cable is wound around each mechanism to raise the road deck 48 m (157 ft) in 12 minutes.

Cathedral Bridge

Historic reference

The piercing form of the Cathedral Bridge's 20 m (66 ft) pylon evokes a scissor or needle, recalling the world of textile manufacturing that made Derby famous. The swinging action of the 60 m (197 ft) long bridge also suggests the cutting action of tailor's shears.

Situated amidst a UNESCO World Heritage Site, the new pedestrian bridge over the River Derwent in Derby required a sensitive and unobtrusive design. The pedestrian and cycle bridge links the eastern bank of the river with the Cathedral Green, the historic town centre and the famous Silk Mill Museum, where the modern factory was born. The new Cathedral Bridge (2009) is a hollow, box steel, cable-stayed pedestrian swing bridge that is designed to rotate swiftly in response to the river's fast-flowing and rapidly rising flood waters. The unobtrusive and slender deck is supported by three steel cables connected to a distinctive pointed pylon.

Pivot

The bridge's swinging mechanism is housed beneath the bridge under the mast at the point where the deck kinks at an angle of 38 degrees. The massive cast steel pivot bearing moves the 90 tonne structure in four minutes.

Swing motion

The counterbalanced bridge is designed to swing very efficiently about a vertical axis. The efficient design allows the bridge to be opened using an electrical motor or operated by hand.

Environmental design

The kinked deck is perfectly balanced to rotate and its thin blade-like profile helps to increase maximum clearance for flood waters. The lean structure also reduces materials, all of which were sourced and fabricated within 24 km (15 miles) of the site.

Stays

Painted mild steel mast

Outward-leaning balustrade

Timber seat

Closed-coil stable stay

Bridge deck

Stainless-steel edge trim

Inward-leaning balustrade

Tail mechanism

Introduction

A cantilever is a structure projecting from a support. In bridge design, the cantilever is one of the four basic structural types. Cantilever bridges are usually balanced on both sides of a support with an anchor (or back) arm forming the back span and opposing the cantilever arm, which forms part of the main span. Compressive forces on the underside of each arm counter the tensile forces in the upper part and are transferred into the support. Most cantilever bridges comprise at least a pair of cantilevers that meet to form the main span.

Cantilever bridges can be constructed from all sorts of materials, using different structures for the cantilever section, such as a beam or a truss. Balancing over a supporting pier, the cantilever arms are either secured to abutments or to adjacent cantilever sections. A suspended section is sometimes inserted between the cantilever arms to extend the main span. To counter the additional vertical loads to the main span, the balancing arms are anchored either to adjacent cantilevers or to solid abutments in the ground at each end of the bridge.

Quebec Bridge
The cantilever bridge with the longest span in the world is the Quebec Bridge (1919) in Canada. The 549 m (1,800 ft) span is due to the extended suspended truss section between two balanced cantilevers.

Forth Rail Bridge

Chords

The upper and lower chords of each cantilever of the Forth Rail Bridge are fabricated from tubular steel and transfer the loads back to the piers. The anchor arms are weighted with 1,000 tonne weights to counterbalance the suspended section and live loads.

Among the most famous cantilever bridges in the world is the railway bridge built over the Firth of Forth in Scotland (1890). The Forth Rail Bridge was the largest execution of the cantilever principle ever attempted in bridge construction. The bridge comprises three enormous balanced double-cantilever sections joined together by small simply supported sections that create two central 521 m (1,709 ft) spans. The two back spans at either end are 201 m (659 ft) long and connect to masonry pylons supporting the approach sections, which are fabricated from steel truss beams.

Human cantilever

A human demonstration was carried out to illustrate the structural theory of the cantilever. The loads are represented by the figure sitting in the middle span on the suspended section. The arms of those seated either side represent the tension in the ties. The wooden struts represent the compression in the lower members. The bricks represent the anchor points at the pylons.

Riveting

This was the first major bridge built entirely of steel. The bridge construction employed more than 4,000 workers a day, 6.5 million steel rivets and 65,000 tonnes of steel. The tubular lower members shown here are made from riveted sections of sheet steel to form a hollow structural tube resisting compressive forces from above and transferring them to the piers.

Forth Rail Bridge *Eiffel Tower*

Grand designs (above)

It would take more than five Eiffel Towers laid end to end to span the same distance as the Forth Rail Bridge's three double-cantilevered sections.

Suspended section

A benefit of the balanced-cantilever method is the extra span gained by inserting a suspended section between each cantilever arm. The suspended sections on the Forth Rail Bridge are 106 m (348 ft) long.

Quebec Bridge

Largest span

Although the cantilever arms of Quebec Bridge are shorter (177 m/581 ft) than the Forth Bridge, the central section is longer, 195 m (640 ft) compared with 106 m (348 ft), making the main span 549 m (1,800 ft).

The Quebec Bridge (1919) in Canada has the longest span of any cantilever bridge in the world. The first attempt to build the bridge ended in catastrophe as the structure collapsed in 1907, forcing a major redesign. The main element of the 987 m (3,238 ft) long riveted steel structure is a through truss. Bearing a resemblance to the earlier Forth Rail Bridge (see pages 180–181), the Quebec Bridge has two balanced-cantilever sections linked by a suspended section forming the main span. The 29 m (95 ft) wide deck carries rail, vehicular and pedestrian traffic.

Centre span

The 195 m (640 ft) central section is a simply supported camel-back truss. The entire central span was hoisted into place between the cantilever arms and connected to the bottom chord. The top chord infill element was installed to complete the structure. The outer arms of the cantilever are anchored to the abutments to counter the loads applied by the weight of the central section and the traffic passing over the bridge.

Cross-bracing (left)

Both side faces of the bridge are tied together by a series of diagonal, vertical and horizontal steel braces. Each of these bracing members is a trussed element. All are elevated high above the deck, which also ties both sides of the bridge together.

K-truss (below)

The three separate structural elements of the bridge (both balanced cantilevers and the central section) are all types of K-trusses. Note how the bottom diagonals pointing back towards the supports are in compression and are further braced to withstand buckling forces.

Montrose Bridge

Suspended section
The Montrose Bridge's central span was extended by a 6.4 m (21 ft) long suspended section, which prevents a catenary (natural) curve forming in the top chord that is characteristic of a suspension bridge.

The curved chords that descend from the towers of the Montrose Bridge (1930–2004) over the South Esk River in Scotland give the impression that this is a suspension bridge. However, on closer inspection the concrete structure is created by two double cantilevers supported on two pairs of concrete piers, forming a main central span with two shorter side spans. The bridge replaced a suspension bridge, using its predecessor's abutments and approaches. The bridge was demolished and replaced in 2004.

Steel reinforcement bar

Concrete pin joint

Chords (above)

The curved top chord contained 76 steel reinforcement bars, each approximately 4 cm (1½ in) in diameter. The high ratio of steel to concrete in a member that is under tension fuelled much debate about whether concrete was necessary at all, and indeed whether it was an appropriate choice of construction material for the structural form with regard to cost and weight.

Concrete sculpture (left)

An angular sculptural concrete pin joint emphasised the meeting of the reinforced concrete piers with the deck and superstructure.

Elevation

The two pairs of towers stood on reinforced concrete piers with 18 m (59 ft) deep piles. The cantilevers, which were formed by concrete trusses with diagonal members between the curved top chord and the deck, created a main span of 66 m (217 ft). The shorter side spans anchored to the abutments were 46 m (151 ft) long.

Story Bridge

Three piers

Uncharacteristically, Story Bridge stands on three piers. At the southern end a main pier supports the cantilever and a secondary pier anchors the structure to prevent rotation. On the northern end the cantilever stands on a main pier but is anchored in the bedrock.

Completed in 1940, the Story Bridge in Brisbane is the largest cantilever bridge in Australia. The steel cantilevers are constructed using a through truss giving a 282 m (925 ft) central span and a clearance of 30 m (98 ft) over the Brisbane River. Construction of the steel structure began with the pylons standing on the concrete piers. The back span arms were erected and anchored to masonry piers and approach roads before the structure cantilevered out into the central span. The inner arms, having reached their maximum length over the river, were joined with a simply supported central section completing the span.

Asymmetrical cantilevers
The outer cantilever arms anchored to masonry piers on land are shorter than the inner arms, which are suspended over the river. This creates an imbalance of load that must be restrained in tension at the anchorage points. The central section joining the two inner arms is a simply supported Pratt truss.

Through truss (above)
The bridge's K-braced steel structure forms a through truss with the road deck sitting on the base of the truss. The structure is supported on masonry piers, whose concrete foundations descend 40 m (130 ft).

Cross-bracing
The outer sides of the bridge are supported directly on the masonry piers and are cross-braced with lateral and diagonal steel trussed elements. The deck is laid above the cross-bracing.

London Bridge

Long history

Over the last two millennia, many bridges have crossed the River Thames near the site of the current London Bridge. The largest of these was a 19-arch, 12th-century bridge supporting tall buildings, a chapel, and a drawbridge.

Opened in 1973, the present 283 m (928 ft) long London Bridge stands near the site of London's earliest bridge, built by the Romans at a time when this part of the River Thames was the tidal limit. The gentle curve of this cantilevered bridge means that it is often mistaken for an arch bridge. However, a number of clues reveal the true nature of this cantilever structure – the two back spans are only half arches and two movement joints are located in the central span. Four box beams fabricated from precast post-tensioned concrete segments form each cantilever section.

Movement joint (below)

Two movement joints are faintly visible in the central span. These indicate the suspended section between the two cantilever arms.

Post-tensioned box beams (above)

Construction was complicated by the dismantling of the previous London Bridge (which was shipped to Arizona, USA), which took place simultaneously while keeping the waterway open to river traffic. The four longitudinal box beams were prefabricated in segments downstream and shipped to the site where they were connected to one another and post-tensioned. Post-tensioning involved inserting steel cables through each segment and tensioning the cables to assist and improve the structural performance of the concrete.

Movement joints

Half arch (below)

The narrower side spans are formed by the outer arms of the cantilever sections, which spring from the slender granite-faced concrete piers. These cantilevers are anchored at the abutments.

Commodore Barry Bridge

Old technology

Commodore Barry
Bridge has the third
longest cantilever
span in the world
and was one of the
last large cantilever
bridges ever built.
Cantilever bridges
have increasingly
been superseded
by cable-stayed
bridges, which are
more economical.

The cantilever section of the 4 km (2½ mile) long
Commodore Barry Bridge (1974) over the Delaware River
in Pennsylvania is the longest cantilever bridge span in the
United States. The main span is 501 m (1,644 ft) long,
with a clearance of 60 m (197 ft), but the cantilever
structure forms part of a longer bridge built of trussed
beams. The downward forces in the cantilever arms that
create the main span are balanced over the central piers
and cause uplift forces in the anchor (or back) span. These
forces are carried by the anchor arm and resisted by the
outer piers and the weight of the back span.

Lateral cross-bracing (left)

Virtually all trussed bridges rely on some form of cross-bracing between their sides. Here 'K'-shaped lateral cross-bracing reinforces the upper part of each section. Similar cross-bracing beneath the 24 m (79 ft) wide deck is strengthened by central longitudinal steel members.

K-shaped lateral cross-bracing

Subdivided Warren truss (right)

The subdivided Warren truss, used here for the approach roads and the main cantilever structure, is defined by the alternating direction of the diagonal members divided by the vertical members. These diagonal members transfer the loads back to the piers.

Subdivided Warren-type through truss

Under and through trusses (below)

The approach roads are simply supported subdivided Warren-type deck trusses (where the road deck sits on the truss beam) that switch to a through truss (where the road deck sits within the truss) at the cantilever section.

Subdivided Warren-type deck truss

Vejle Fjord Bridge

Expansion

The Vejle Fjord Bridge is divided into four sections to allow for the expansion of the prestressed concrete box girders. Each expansion section is about 490 m (1,610 ft) long.

The 1,712 m (5,617 ft) long bridge over the Vejle Fjord (1980) in Denmark has 15 spans of approximately 110 m (360 ft). It was constructed using the free cantilever method. Each double cantilever was constructed over the piers and built out to meet at the centre span. The joint between each 'T' is visible in the centre of the span. In its permanent form, the bridge is a haunched concrete box girder – the underside arcs from the support to the mid-span – which transfers load back to the piers.

Slender geometry (right)

The construction of a free cantilever requires the structure to have greater depth close to the supports getting shallower towards the outer ends of the cantilever arms, creating a slender section at mid-span. The narrow piers support the haunched box-girder sections that arc from a depth of 6 m (20 ft) over the piers to a shallow 2.5 m (8 ft) in the centre of the span. The maximum clearance is approximately 40 m (130 ft).

Concrete piers in direction of span

Perpendicular piers (above)

To maintain the bridge's slim appearance, the concrete piers have been designed to be shallow in the direction of the span, which is made possible by the balanced cantilever form minimising bending forces. The breadth of the piers offers lateral stability to the suspended deck.

Cantilever deck

The principal structural elements, the pier and central box girder, are the same width and support the wider deck. The deck is a cantilevered concrete slab that projects beyond the sides of the box girder on tapering prestressed transverse beams. The full width of the thin concrete deck is 29 m (95 ft).

Introduction

The deck of a suspension bridge is hung from suspended cables in tension, which transfer the loads to anchorages at either end of the bridge or to towers where they are carried to the ground. Traditional (or simple) suspension bridges comprise a deck between two suspended cables hung between two fixed points. Modern suspension bridges adopt the same principle but use two towers over which the cables are strung to create one main central span and two smaller side spans. The cables of a modern suspension bridge are typically made from a single strand wound continuously from one side of the bridge to the other. At each pass, it is fixed into anchorage blocks embedded deep in the ground. These anchorages are designed to resist the massive tensile forces in the cables that are generated by carrying the dynamic loads through the vertical hangers to the deck. Most of the weight, which combines the structure and the live loads travelling over it, is carried as tensile forces through the hangers and cables and borne by the tower, which is placed under compression. The structural efficiency of suspension-bridge design makes it the most effective bridge system for crossing long distances with a single span.

Akashi Kaikyo
The longest suspension bridge in the world is the Akashi Kaikyo Bridge (1998) in Japan, which measures just under 4 km (2½ miles) and has a main span of 2 km (1¼ miles).

Menai Straits Bridge

Rust prevention
The suspension chains of the Menai Straits Bridge were the longest ever constructed. Between fabrication and installation on site, each link of iron chain was stored in warm linseed oil to prevent rusting.

Linking mainland Wales and the island of Anglesey across the treacherous waters of the Menai Straits is one of the earliest modern suspension bridges and, in its time, the longest span in the world. The 305 m (1,000 ft) long bridge has a 176.5 m (579 ft) span and stands 30 m (98 ft) above sea level. Construction of the Menai Straits Bridge (1826) began in 1820 with the erection of the masonry piers, followed by the wrought iron suspension chains supporting a double-carriageway timber deck that was laid over iron beams.

Masonry approaches (left)

Limestone piers support the approach roads and comprise four 16 m (52 ft) wide round arches on one side of the Straits and three on the other. The arched approaches culminate at the 45.7 m (150 ft) high marble-faced towers over which the suspension chains are hung.

Trussed parapets (right)

The oak deck often succumbed to strong winds and was replaced by a steel deck in 1893, increasing its dead weight from 633 tonnes to 1,016 tonnes. In 1938, the iron chains, rods and steel deck were replaced by steel elements. Longitudinal trussed parapets on the outer sides of the carriageway support the deck and were concealed by cantilevered pedestrian walkways.

Iron chains

Longitudinal truss

Steel deck

Suspension chains (left)

The deck was originally supported by four sets of primary suspension chains secured to one another through eyebars at each end. Each primary chain weighed 121 tonnes and comprised 935 wrought iron sections in four individual chains. These were hauled across the river by 150 workers and secured into 18 m (59 ft) deep anchorages tunnelled into the bedrock and clamped in place with 3 m (10 ft) long iron bolts.

Clifton Suspension Bridge

SUSPENSION BRIDGES

Endurance

When the Clifton Suspension Bridge was built, the heaviest loads it was designed to carry were those of a horse and cart. The same structure continues to carry thousands of motorised vehicles over it every day.

The Clifton Suspension Bridge (1864) over the River Avon in Bristol, was the first commission for the young engineer, Isambard Kingdom Brunel. Construction started in 1831 but the project was beset with difficulties and eventually stalled. It was not completed until five years after Brunel's death. The total length of the bridge is 414 m (1,358 ft). The wrought iron chains sit in saddles 4 m (13 ft) below the top of the piers and descend 21 m (69 ft), giving the 9.5 m (31 ft) wide deck a 1 m (3 ft) camber and clearance above the river of 76 m (249 ft).

Iron links (left)

The suspension chains are made from wrought iron links bolted together every few metres rather than wire bundled into cables, which later became standard. Each section of chain comprises ten or eleven flat iron links stacked in three independent vertical layers. Some of the chain links came from the dismantled Hungerford Bridge in London, also designed by Brunel.

Suspension rods (below)

The main span of 214 m (702 ft) comprises a deck suspended from 81 pairs of wrought iron suspension rods. Ranging in length from 1 to 20 m (3 to 66 ft), these rods are arranged vertically and secured at the top end to the main chains and at the bottom end to the road deck.

Pylons (above)

The elliptically arched pylons, each 26 m (85 ft) high, are constructed of Pennant stone and stand on sandstone abutments. Each pylon houses roller-mounted saddles over which the chains are hung, allowing for longitudinal movement in response to shifting loads and temperature variations, and to prevent transverse movement.

John A. Roebling Bridge

Suspension cables
The John A. Roebling Bridge had more than 10,000 wires woven between the piers to create the two main suspension cables supporting the deck with a clearance of 30 m (98 ft).

Posthumously renamed after its designer, the John A. Roebling Bridge over the Ohio River on the border of Kentucky and Ohio, USA, was the longest suspension bridge in the world when it opened to the public in 1866. With prevailing knowledge and expertise of suspension-bridge design, Roebling understood that the Ohio River was too wide for a suspension bridge to cross from shore to shore, so he designed a bridge with two masonry towers standing in the river 300 m (980 ft) apart.

Suspender cable

Stiffening truss

Old deck design

New deck design

Cable stays

Deck cross section

Deck structure

The deck is supported by iron suspender cables that hang from the main cables. Adjacent pairs of suspender cables are connected by wrought iron beams, over which the deck is laid. Each deck section between lateral beams is strengthened by longitudinally arranged members and diagonal and transverse bracing, which provide lateral stiffness.

Widened deck

By the end of the 19th century, the deck had been widened and an extra pair of suspensions cables laid over the towers to increase its load capacity. The original deck surface is relatively thin with cable stays beneath the deck and longitudinal trusses for stiffness. The new 18 m (59 ft) wide deck is strengthened and stiffened by rigid and wire types of cross-bracing.

Masonry piers

The foundations of the 23 m (75 ft) high limestone-faced sandstone piers sit on 13 perpendicularly arranged layers of oak timbers bolted together and encased in concrete. Originally, the piers were designed to carry a much larger structure than was eventually built owing to the disruption of the American Civil War.

Brooklyn Bridge

The Brooklyn Bridge (1883) in New York, USA, was the longest suspension bridge in the world when completed. To span such a wide expanse of water, four suspension cables were used instead of two. These were strung over saddles in both masonry pylons, above the iconic pair of pointed arches. Secondary cables radiating out from the pylons give the bridge its distinctive fan-like appearance, making it a hybrid of the cable-stayed and suspension systems for the portion close to the pylons.

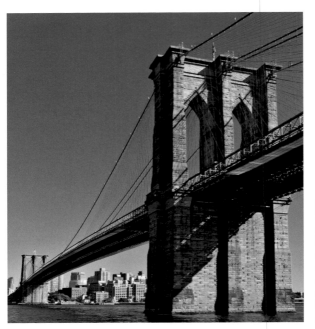

Steel cables

The Brooklyn Bridge was the first suspension bridge to use steel-wire cables. Previously, suspension bridges used iron cables or chains, which were heavier and less structurally efficient.

Cable anchors

The four main cables were spun from 5,434 strands of continuously wound wire strung back and forth across saddles in the bridge then tightly bound together and anchored at either shore. Each of the four cables could carry 11,200 tonnes. The vertical cables, or hangers, were manufactured from more than 23,000 km (14,300 miles) of wires, intertwined like a rope.

Overhead view

Deck cross section

Elevated walkway · Diagonal steels · Elevated railway

Traffic lane · Streetcar lane · Streetcar lane · Traffic lane · Steel truss

Transverse truss beams

Deck section (above)

The deck originally carried two lanes of elevated railway, two lanes for streetcars and two lanes of vehicular traffic, with an elevated pedestrian walkway in the middle. Longitudinal steel trusses on the outside of the deck and larger ones separating the railway from the outer lanes support the deck. The base of the deck comprises transverse truss beams spanning beneath the longitudinal trusses. The deck is braced with diagonal steels to give lateral stability.

Elevation (below)

The entire length of the bridge is nearly 2 km (1¼ miles), with a main span between the pylons of 486 m (1,594 ft) and a maximum clearance of 41 m (134 ft). The 84 m (276 ft) high pylons once towered over the New York skyline.

George Washington Bridge

Busiest bridge
More than five million vehicles crossed the George Washington Bridge in its first year. Today, nearly 100 million vehicles cross the bridge each year, making it the busiest bridge in the world.

The George Washington Bridge (1931) over the Hudson River in New York, USA, was revolutionary in its time, with a main span more than twice as long as any other in the world. It was constructed with just one deck, but designed to support two. A second deck was later added and opened in 1962. The engineer, Othmar Ammann, knew that the weight of the deck could be used to stabilise the bridge, eliminating the need for stiffening trusses which were common in earlier suspension bridges. Each of the bridge's 32 m (105 ft) longitudinal deck beams weighs 66 tonnes. Four main steel suspension cables, each nearly 1 m (3 ft) in diameter, carry the 36 m (118 ft) wide twin decks across the 1,067 m (3,500 ft) span between towers. The length of the bridge between abutments is 1,451 m (4,760 ft), with a clearance of 65 m (213 ft).

Additional deck

The additional lower deck was constructed beneath the existing deck using 8 m (26 ft) deep steel subdivided Warren-type trusses. The addition of a deck forms a structural extension, since the top deck and lower deck are acting as a single structural entity rather than one being hung from the other.

Wire cables

Each of the four main cables comprises a single strand wound back and forth across the river 61 times and is secured in anchorage blocks that are made from 260,000 tonnes of concrete. Each strand is fabricated from 434 individual wires with a total length of 172,000 km (107,000 miles).

Trussed towers

Each of the two steel towers supporting the four suspension cables is 183 m (600 ft) high and weighs more than 21,000 tonnes. The towers were originally intended to be clad in concrete and granite, but the appealing functional appearance of the exposed steel truss skeleton resulted in them remaining unclad, also saving money.

Golden Gate Bridge

Distinctive profile
The towers of the Golden Gate Bridge, which soar to a height of 227 m (746 ft) are made up of hollow steel cells riveted together. After the towers were completed the cables were spun back and forth across the gulf.

The iconic structure spanning the Golden Gate (1937) at the mouth of San Francisco Bay, USA, as it opens out into the Pacific Ocean is one of the world's most famous bridges. The 27 m (89 ft) wide deck is suspended from 250 pairs of suspender cables hanging from a pair of main cables. Each main cable is 1 m (3 ft) in diameter and 2.332 m (7,650 ft) long, and fabricated from 27,572 strands of galvanised steel wire with a total length of 129,000 km (80,000 miles). Securing these massive cables and their tensile loads requires a concrete anchorage weighing 60,000 tonnes at either end of the structure.

Environmental effects (below)

Temperature changes cause the deck to expand and contract, which affects the deck's length and, consequently, its height above water. The Golden Gate's vertical variation is as much as 5 m (16 ft) offering a maximum clearance of 67 m (220 ft). The deck is designed to reduce wind resistance and constructed using cross-braced, steel subdivided Warren-type trusses, allowing the wind to pass through with minimal resistance.

Styling (above)

The style of a bridge often reflects the era in which it was built. Few bridges are more evocative of their time than the Golden Gate Bridge with its tapering Art Deco towers, whose horizontal bracing forms a Vierendeel truss. Each tower stands 227 m (746 ft) tall and weighs 44,000 tonnes. The bridge's distinctive orange-vermillion colour was chosen to complement the setting sun over the Pacific.

Longest span (below)

The main span of 1,280 m (4,200 ft) broke all records when completed. The total length of the suspended structure, including the two 343 m (1,125 ft) side spans, is 1,966 m (6,450 ft).

Lions Gate Bridge

Elevation

The 1,823 m (5,980 ft) long Lions Gate Bridge comprises four sections: a main span of 472 m (1,548 ft), two side spans of 187 m (613 ft), and a 670 m (2,200 ft) approach road.

Originally opened as a two-lane road bridge in 1938, the Lions Gate Bridge near Vancouver, Canada, was unable to cope with the growing volume of traffic. It was decided to replace the deck while keeping the bridge open. The road deck was replaced in sections through nights and weekends between 2000 and 2001. Improvements in bridge design allowed the existing steel towers and cables to carry the new deck which, although 35 percent wider than the original deck, weighed the same.

Replacing the decks (above)

The original 12 m (40 ft) wide deck contained two lanes of traffic and two pedestrian walkways inside longitudinal trusses that formed the sides of a 'U'-shaped structural deck. The new 17 m (55 ft) wide deck had a central section of 12 m (39 ft), from which 2 m (6½ ft) wide pedestrian and cycle lanes are suspended on cantilevered sections either side. The trusses that once restricted views from the deck were replaced by a stiffening truss beneath the deck.

Tower trusses (right)

The towers are strengthened by a combination of lateral supports – diagonal cross-bracing and narrower tapered Vierendeel trusses two-thirds up the tower and also at its apex. A horizontal beam between the uprights at the deck level also supports the deck.

Anchorage

The main cables are anchored differently at either end. One is secured into the bedrock where the deck meets the valley side and the other is saddled over the pier separating the approach ramp from the suspended section and taken down to an anchorage on the valley floor.

Verrazano Narrows Bridge

Expansion
Verrazano Narrows Bridge's location exposes it to severe climatic conditions, causing it to close in winter or during high winds. Thermal expansion causes the deck elevation to change by 3.5 m (11½ ft) between winter and summer.

The longest central span of any bridge in America is the suspension bridge across the Hudson River between Staten Island and Brooklyn. With a span of 1,298 m (4,258 ft), the Verrazano Narrows Bridge was not only the longest bridge in the world when completed in 1964, but it also supported a double carriageway. In order to carry the additional load the bridge has two pairs of suspension cables rather than the more common single pair. The deck truss receives additional support by sitting on cross beams between the uprights of each tower.

Double deck (left)

Fabricated from 60 steel truss sections weighing 400 tonnes apiece, the double deck is supported from its upper surface. The lower deck is suspended below and comprises a cross-braced 5 m (16 ft) deep subdivided Warren-type through truss, which provides lateral stability.

Unparallel towers (right)

Standing 298 m (978 ft) apart and 211 m (692 ft) high, the design of the steel towers, which stand on 51 m (167 ft) deep foundations, had to accommodate the curvature of the earth – their tops being 41 mm (1²⁄₃ in) farther apart than their bases.

Twin cables

Each side of the bridge is supported by a pair of cables, each one made up of 26,108 strands and 230,000 km (143,000 miles) long. The cables weigh 9,270 tonnes and measure almost 1 m (3 ft) in diameter. The strands were laid in pairs by spinning wheels that passed back and forth between the anchorages, and are clamped along the length of the cable.

Humber Bridge

Completed in 1981, the road bridge over the River Humber was once the longest suspension bridge in the world. The previous longest span of a suspension bridge with concrete towers was 608 m (1,995 ft), but the Humber Bridge's main span was 1,410 m (4,625 ft). The technological innovation that allowed this significant increase was the novel use of hollow reinforced concrete sections in the 155 m (508 ft) high towers. This allowed the towers to be lighter and taller, reducing the load on the foundations.

Environmentally responsive
The decision to build a suspension bridge was influenced by the environmental conditions in the River Humber, which are characterised by a shifting estuary bed. Shipping channels are therefore constantly changing and only a suspension bridge, which uses only two piers to create a large span, was suitable.

Hollow concrete towers (below)

The towers were constructed using slip-forming technique, in which the concrete is poured in formwork that rises with the structure, in the same manner of construction as some skyscraper cores. The two slightly tapered pylons in each tower are horizontally cross-braced to form a Vierendeel truss for extra stability.

Steel cables (left)

Each of the two main steel cables is 70 cm (27½ in) in diameter and comprises 37 strands fabricated from 404 lengths of 5 mm (⅕ in) diameter high-tensile wire. Every strand is tied and anchored at either end of the bridge before being spun into cables. Each cable weighs 5,500 tonnes and each reinforced concrete anchorage weighs 300,000 tonnes.

Aerodynamic deck (above)

The 28 m (92 ft) wide and 4.5 m (14¾ ft) deep deck is fabricated from 18 m (59 ft) long 140 tonne steel hollow box sections. These sections were fabricated off-site and suspended from hanger cables before being welded together. The materials and design reduce the deck's weight and its razor sharp aerodynamic profile increases its stability by minimising the lateral forces created by wind loads.

Tsing Ma Bridge

Side spans

The cables over each of the Tsing Ma Bridge side spans are anchored at the ground rather than at deck level. One set is straight rather than curved as it carries no vertical load.

The world's longest suspension bridge carrying both rail and road traffic is the Tsing Ma Bridge (1997) in Hong Kong, China. The 2,160 m (7,087 ft) long bridge carries a double deck with six lanes of traffic above two railway lines and two emergency traffic lanes. The asymmetrical structure comprises three sections: a main span, a side span partly suspended and partly supported on concrete piers, and an approach road supported on concrete piers.

Contained double deck (right)

The depth of the 49,000 tonne double deck exposes it to high winds in an area prone to extreme weather. The deck has been designed to minimise wind resistance. Its aerodynamic edge profile encourages cross winds to pass around it while a continuous opening along the middle of the upper deck prevents air pressure building up within the deck.

Bridge span (above)

The two main cables are strung over saddles at the top of the 206 m (676 ft) high and 52,000 tonne reinforced concrete towers. The side spans comprise a partially suspended section and an approach road supported on three evenly spaced reinforced concrete piers 72 m (236 ft) apart.

Anchorage blocks (below)

The two main cables are made from 33,400 steel wires each 5 mm (⅕ in) in diameter. Weighing a total of 26,700 tonnes, the two cables support a load of 106,000 tonnes and are secured at either end in anchorages weighing 200,000 and 250,000 tonnes respectively. The face of these concrete structures is angled so that it is perpendicular to the line of the tensile forces in the cables.

Great Belt Bridge

Suspended section
Suspension bridges are ideal in busy shipping areas as they provide unobstructed clearances. The main span of the suspended section of the Great Belt Bridge provides a 1,624 m (5,328 ft) wide navigation channel.

The suspended section of the Great Belt Bridge (1998) forms part of a 14 km (8¾ mile) long parallel road and rail bridge (and tunnel) system across the strait between the Danish islands of Zealand and Funen. Most of the bridge comprises a continuous concrete beam supported on reinforced concrete piers. An interesting feature of the bridge is that the suspension cable anchorages are not land based. Massive concrete sea anchorages resist the tensile forces with profiled wedges providing a frictional resistance beneath the seabed.

Anchor cross section

Caisson

12 m
(39 ft)

Gravel
wedges

22 m
(72 ft)

Artificial
island

Rear

Front

12.5 m (41 ft)

Sea anchorage (above)
The anchorage's hollow shape reduces the
amount of concrete without compromising
performance, making them lighter
structures, aesthetically and physically.
The shape reflects the tensile forces the
anchorage is resisting. Each anchorage
is 122 m (400 ft) long at their base with
a foundation 22 m (72 ft) beneath the
seabed and angled to resist the horizontal
loads pulling them inwards.

Deck section
The 31 m (102 ft) wide and 4 m (13 ft)
deep aerodynamic welded steel box-girder
deck is designed to resist wind loads.
Transverse trusses inside the box section
increase the deck's rigidity.

Continuous suspension (below)
The deck is suspended continuously
on two 3,079 m (10,100 ft) long
cables between anchor points 2.7 km
(1⅔ miles) apart. This design feature
causes the 254 m (833 ft) high
reinforced concrete towers to be higher
than normal. The towers' tapered form
reflects the compressive forces they are
under, tapering from their solid bases
and separating 21 m (69 ft) above sea
level into two pylons cross-braced
midway and at their top.

Deck cross section

Transverse
truss

Akashi Kaikyo Bridge

Earthquake proof
The 1995 Kobe earthquake moved the towers of the Akashi Kaikyo Bridge apart during construction, causing the main span to increase by 1 m (3 ft). Hinges in deck girders are designed to increase resistance to wind and earthquakes.

Crossing the Akashi Strait between Kobe and Iwaya in Japan is the longest span suspension bridge in the world. The 3,911 m (12,831 ft) long deck is suspended 97 m (318 ft) above the sea by two huge steel cables over 1 m (3 ft) in diameter, the 36,830 strands of which would wrap around the world seven times. Construction began in 1988 and took ten years to complete, consuming 181,000 tonnes of steel and 1.4 million m^3 (50,000 cu ft) of concrete.

World's longest span (below)

The cables are hung over a pair of 297 m (974 ft) high pylons that create three spans: a central span of 1,991 m (6,532 ft) and two outer spans of 960 m (3,151 ft). Note how the hangers connecting the deck to the suspension cables are fixed above each vertical truss member. If they were fixed midway between vertical members then each member would also take bending moments.

Concrete anchorage (above)

With such a large span generating large tensile forces, the cables have to be secured by huge anchorages. Each one is constructed using 350,000 tonnes of concrete and has been combined with the abutments. Note how the approach road is a concrete beam supported on the abutments, beyond which the deck becomes a subdivided Warren-type steel trussed box.

Elevator inside tower

Main tower

Caissons

Cable

Cross-bracing

Deck

Concrete foundation

Foundations

The bases of each pylon are set into deep circular foundations constructed with concrete and steel to provide a solid base for the towers. Each tower is cross-braced for lateral stability but remains relatively open to minimise wind forces.

Si Du River Bridge

Asymmetry
One side of the
Si Du River Bridge
comprises an
approach road,
whereas the deck
on the other side
meets the valley wall
at the base of the
tower, creating an
asymmetrical profile.

Crossing the valley of the Si Du River in Hubei Province, China, at a height of approximately 500 m (1,640 ft), is the world's highest suspension bridge. The 1,365 m (4,478 ft) long bridge opened to traffic in November 2009. The bridge has a main span of 900 m (2,952 ft) but there are no suspended side spans because of the constricted geological setting. The main suspension cables are strung over the 'H'-shaped towers and then anchored directly into bedrock behind.

Deck cross section

Truss deck (left)

The 26 m (85 ft) wide deck is a frame made up of a series of prefabricated steel trusses. Longitudinal subdivided Warren-type trusses run along the outside of the entire length of the deck. These are cross-braced with regular and evenly spaced transverse trusses that stiffen the structure.

Non-horizontal deck (right)

The 'H'-shaped reinforced concrete pylons on each side of the valley differ in height by 5 m (16 ft), creating a slight gradient in the road deck. Note how the main cables behind the towers have no hangers as they carry no vertical loads and serve only to anchor the tensile forces in the main cable. The cables therefore form a straight line behind each tower but a catenary curve across the main span.

Rocket-launched cable

The bridge is supported by two cables, which were spun continuously back and forth across the valley. The first pilot cable, rather than being carried down and across the valley, was fired across by a rocket.

Introduction

Cable-stayed bridges are a type of suspension bridge, but they differ fundamentally from conventional suspension bridges. They share the same principle of using a series of tensile cables to transfer loads from the deck to a tower, which is placed under compression, but instead of a continuous cable, the cables of cable-stayed bridges are independent of one another and support deck sections extending out from the tower. This makes it easy to measure the forces in each cable and adjust its length as necessary. It also makes it easier to maintain the bridge, as individual cables can be replaced without compromising the bridge's overall strength. Cable-stayed bridges contain the loads within their structure, thus eliminating the need for an anchorage. This makes them particularly well suited to situations in which anchorages are not available, such as multiple spans or where the geological conditions provide an inadequate substrate. The deck load is transferred to the tower by symmetrical sets of cables, which balance the structure. Other types of cable-stayed bridges use inclined towers to balance the deck loads.

Rio–Antirio Bridge
Nearly 3 km (2 miles) long, with a suspended deck of more than 2 km (1¼ miles), the Rio–Antirio Bridge in Greece is the longest multi-span cable-stayed bridge in the world.

Skarnsund Bridge

High-strength concrete

The slender profile of the Skarnsund Bridge's deck and piers was achieved through recent advances in concrete technology that exploit the high-strength characteristics of modern concrete.

The Skarnsund Bridge (1991) in Norway was among the earliest examples of large cable-stayed bridges. With a total length of 1,010 m (3,314 ft), the bridge is a twin-pylon structure with three spans – one main span and two different-sized outer spans. The main span is 530 m (1,740 ft) long and 45 m (147 ft) above the water. One side span is connected directly to approach roads on the cliff and the other is partially supported by four concrete piers to the approach roads further inland. The 13 m (43 ft) wide and 2 m (6½ ft) deep, box-girder deck is the longest concrete cable-stayed deck in the world.

Cable anchorage (left)

Twenty-three pairs of cables weighing a total of 1.030 tonnes attach to each side of the mast at the apex of each tower. The cables vary in diameter from 52–85 mm (2–3⅓ in).

Triangular tower (right)

The A-frame shape of each 152 m (500 ft) tall tower is designed for its inherent lateral stability, being wide at its base and narrow at its top. The tensile forces are gathered at the top of the vertical mast and transferred down the two legs to the base of the tower and into the foundations.

Deck

Flange

Cross beam

Deck support

The load from the deck is supported both by the cables and the cross beam between the legs of the tower. Small flanges on the side of the deck close to the tower are designed to prevent significant longitudinal movement in the deck. The deck's thin aerodynamic triangular section makes it lightweight and structurally stable.

Pont de Normandie

Continuous beam

The approach ramp of the Pont de Normandie is made from a continuous concrete beam supported on piers that rise incrementally from the shore to the A-frame tower.

Cable-stayed bridge design has developed rapidly in recent years. A demonstration of this progress is the Pont de Normandie, which marked a revolution in cable-stayed bridge design when constructed. When it opened in 1995, it was the world's longest cable-stayed bridge (2,143 m/ 7,030 ft) and had the world's longest cable-stayed span (856 m/2,808 ft), but the bridge has since been surpassed by many larger and longer cable-stayed bridges.

Cable connectors (left)

The deck is supported every 19.6 m (64⅓ ft) by 23 pairs of steel cables on either side of the tower. These splay in a fan-like arrangement from the apex of the tower where they are tied into steel sleeves embedded into a steel box, which is built into the mast. These transfer the tensile forces in the cables into compression forces in the towers.

Approach ramp

The last section of the approach ramp spans over the cross beam in the tower and connects to the lighter steel box girder in the bridge's main span. The deck's slender profile is designed to minimise lateral forces generated by cross winds.

Approach ramp fabricated from a continuous concrete beam

A-frame tower (right)

The overall strength of the bridge derives from the structural efficiency of the 'A'-shaped frame of the boxed reinforced concrete pylons, which transfers the loads from the deck to the ground through its legs. The two legs provide a solid base and the horizontal beam supports and laterally restrains the deck. The cables are fixed from the vertical section of the mast above the point where the two legs join.

Vasco da Gama Bridge

Tied piers

The Vasco da Gama Bridge's legs taper outwards below the cross beam to the base and support the compressive forces from above. The legs are tied at their bases by concrete foundations that protect the piers from accidental impact from ships.

The 826 m (2,710 ft) cable-stayed section of the Vasco da Gama Bridge (1998) forms a small part of the 17 km (10½ mile) long bridge across the Tagus River in Portugal. The main cable-stayed span is 420 m (1,377 ft) long and the side spans are 203 m (666 ft). The 30 m (98 ft) wide steel box-girder deck of the six-lane motorway is carried across the three spans by 192 cables suspended from the two 'H'-shaped pylons. The inward-leaning lamp posts on the deck are specially designed not to disturb the marine environment by shining directly on the water.

Deck support (left)

In the absence of any support beneath the deck, two cables either side of the road deck on the inner face of the tower support the deck as it passes through the tower.

H-frame towers (right)

The cables are secured to the upper vertical portions of the 155 m (508 ft) tall H-frames and attached to the outer edge of the road deck. The tower structure is stiffened by the large, slightly haunched cross beam above the road deck. Note there is no cross beam at the head and none at deck level, meaning the deck is suspended throughout (the deck does, however, appear to be fixed to the legs for lateral restraint). The foundations of the towers' legs form a protective cross beam visible above the water level.

Twin cables

Some 184 cables of varying thickness radiate from the four vertical masts in a harp-like arrangement. In a true harp-like pattern, each cable stay is parallel to the next. Here they are not quite parallel, but neither do they radiate from a single compressed point as in the case of a fan-like pattern.

Millau Viaduct

Piers

Each of the Millau Viaduct's piers tapers from 27 m (89 ft) at the base to 10 m (33 ft) at the deck, creating a form that reduces its presence in the landscape.

Bridging the wide valley created by the River Tarn in southern France, the Millau Viaduct (2004) is the longest cable-stayed bridge (2.5 km/1½ miles) and the tallest bridge (343 m/1,125 ft) in the world. The structure is divided into eight spans created by seven piers supporting 90 m (295 ft) high masts. Eleven pairs of cables on each mast support the 36,000 tonne steel road deck, which has a gradient from north to south of 3 per cent.

Longest cable-stayed deck (above)

The equally spaced piers range in height from 77–244 m (253–800 ft), creating six central spans of 342 m (1,122 ft) and two outer spans of 204 m (669 ft). The cables are arranged in a radial formation fixed at equally spaced anchorages on the road deck and up the pylons. The upper part of the pylon, above the cable anchors, serves no structural purpose.

Deck cross section

Curved Parapets

Outer triangular deck section

Steel box-girder spine

Trapezoidal steel bracing

Mast cross section

Mast side view

Road deck

Pier splices into two

Road deck (above)

The 32 m (105 ft) wide road deck comprises three parts: a central spine fabricated from a 4 m (13 ft) wide and 4.2 m (13¾ ft) deep steel box girder; hollow outer deck sections fabricated from triangular elements welded to the spine; and trapezoidal steel bracing inside the deck to stiffen it. The cables connect to the central box girder spine. The curved three metre high parapets significantly reduce wind loads.

Masts (right)

Each pier splices into two beneath the road deck to increase the structure's longitudinal flexibility in order to accommodate the road deck's expansion and contraction. The 87 m 285 ft) high and 700 tonne A-frame masts above deck mirror this structural response and can be seen in the right-hand example of the elevations.

Charilaos Trikoupis (Rio–Antirio) Bridge

Alternative design
The distance of the strait favoured a suspension bridge, but geological and seismic considerations forced an alternative design approach for the Rio–Antirio Bridge, leading to a multi-span cable-stayed bridge.

The 2,880 m (9,449 ft) long Rio–Antirio Bridge (officially Charilaos Trikoupis Bridge, 2004) connects the Peloponnese to mainland Greece. It comprises approach roads and three main spans and two side spans, which are 560 m (1,837 ft) and 286 m (938 ft) long respectively. Four pylons, each with two sets of cables comprising 23 pairs on both sides of the deck, support the 2,252 m (7,388 ft) long road deck. The deck is a composite cross-braced steel frame structure of longitudinal I-beams transversely braced with steel beams and strengthened by concrete slabs between.

Elevation (above)

The cable-stayed bridge section is suspended from four towers. Expansion joints at each end of the suspended section as well as the deck's structural independence from the four piers (at deck level) allow the deck to respond flexibly and safely to climatic and seismic conditions by accommodating longitudinal deck movements of up to 5 m (16 ft).

Foundations (below)

In the absence of bedrock at an appropriate foundation level, the 65 m (213 ft) tall pylons sit on massive radial reinforced concrete foundations 90 m (295 ft) in diameter embedded in the soil and supported on friction piles in the seabed.

Square towers (above)

The reinforced concrete towers are designed to transfer the loads from the deck directly and efficiently to the piers. The tensile forces in the cables are gathered in the mast and transferred down the four legs of each tower to the corner of the capital of each pier. The four-legged configuration provides additional stability in the event of an earthquake. The capital transfers the compressive loads into the octagonal piers and into the foundations.

Sutong Bridge

Triangular pylons
Sutong Bridge's inverted 'Y'-shape legs are connected by post-tensioned tie beams beneath the deck and designed to resist the impact of a 50,000 tonne ship. The cross beam ties the pylon legs and resists the outward-thrust forces.

The navigation channel in the 8.2 km (5 mile) long Sutong Bridge (2008) across the Yangtze River in China is created by the longest cable-stayed bridge in the world. Two sets of cable stays (one on each side of the deck) connect the pair of 300 m (980 ft) high pylons to the steel box-girder deck to create a main span of 1,088 m (3,570 ft). With the bedrock 270 m (885 ft) beneath the river, the pylons and approach piers are supported in the overlying soils by hundreds of friction piles bored to a depth of around 100 m (330 ft).

Deck cross section

Transverse plates — Road deck

Stay cable

Box-girder deck

The cable-stayed deck section is a 41 m
(134 ft) wide and 4 m (13 ft) deep steel
box girder fabricated from 16 m (52 ft)
long 450 tonne segments. Each segment
was hoisted from barges and welded
together. Longitudinal closed steel troughs
and transverse plates every 4 m (13 ft)
along the box section stiffen the deck,
with the transverse plates increasing
in frequency nearer the pylons.

Cable connection

The cables fan out from the upper
portion of the pylon meeting the centre
span at regular 16 m (52 ft) intervals and
the back span at 12 m (39 ft) intervals.
The deck was constructed in sections and
the individual cables were adjusted as each
deck section was laid. The longest cable to
the mid-span is 577 m (1,893 ft) long and
weighs 59 tonnes.

Approach

The approach roads are
fabricated from hollow,
single-cell concrete sections
forming a continuous girder
over transverse concrete piers.

Hangzhou Bay Bridge

Coastal exposure

Hangzhou Bay Bridge's potential exposure to typhoon winds, earthquakes and extreme tides complicated both the design and construction process, which took more than a decade to complete.

The two cable-stayed sections of the 36 km (22 mile) long Hangzhou Bay Bridge (2008) in eastern China provide navigation channels in what was (until 2010) the longest seaborne bridge. The cable-stayed sections of the 37 m (121 ft) wide, streamlined road deck are made from 15 m (49 ft) long and 3.5 m (11½ ft) deep steel box-girder sections. These are attached to approach roads constructed from concrete box-girder sections that form a continuous beam supported on pairs of concrete piers.

North and south sections (right and below)

The south channel comprises a two-span A-frame single pylon. The three-span north channel comprises a pair of diamond-shaped pylons. The main spans of the north and south channels are 448 m (1,470 ft) and 318 m (1,043 ft) respectively. Each pylon supports a double set of cables fixed into anchor plates on adjacent sides of the road deck.

South section

A-frame single pylon

North section

Diamond-shaped pylon

Through pylons and deck seating (right)

The pylons are designed to provide structural stiffness and maximise resistance to wind forces. In each case the road deck passes through the pylon's frame, which is diamond-shaped in the north channel and 'A'-shaped in the south channel. The cross beam at the diamond centre resists the outward-thrust forces generated by the inclined piers above. As it passes through the north and south pylons the road deck is restrained by cross beams, which provide lateral stability.

Stonecutters Bridge

Crossing the busy Rambler Channel in Hong Kong, China, the Stonecutters Bridge (2009) comprises a 1,596 m (5,236 ft) long, twin-pylon, cable-stayed structure with a main span of 1,018 m (3,340 ft) – the second longest cable-stayed span in the world. The two 298 m (978 ft) high needle towers support the main deck with 27 pairs of cables extending from both sides of each needle tower. The cables are anchored to the main span at 18 m (59 ft) intervals and at 10 m (33 ft) intervals to the side spans.

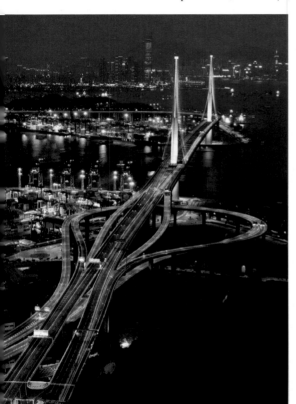

Approach

The 51 m (167 ft) wide and 4 m (13 ft) deep twin deck separates as it approaches each tower with three lanes of traffic on either side. Strength is maintained by cross girders that tie the twin decks together. These can be seen between the decks in the mid-span section.

Needle tower (below)

The tapering form of the needle towers from the broad base to the pointed apex reflects the structural forces as the pylon cantilevers up from its foundations. The first 175 m (574 ft) is constructed from reinforced concrete, above which the concrete core is surrounded by a skin made of 32 stainless-steel sections. The composite design reduces oscillation caused by resonance in the tower and in the cables.

Cable connections (above)

The bottom three sets of cables are set directly into the concrete pylon core. All other cables are fixed into steel anchor boxes. At the top of each cable, 16 mm ($\frac{2}{3}$ in) diameter and 300 mm ($11\frac{1}{5}$ in) long stainless-steel shear connectors transfer the load between the concrete core, steel skin and anchor box.

Cross girder   Concrete core

Tower cross section

Twin deck

The deck of each side span is fabricated from a continuous concrete box-girder beam supported on four concrete piers. The main span comprises a lightweight aerodynamic steel box girder that minimises dead loads. The deck is strengthened and stiffened internally by transverse and longitudinal fin plates. Strength is maintained by cross girders that tie the twin decks together.

Jiaozhou Bay (Qingdao Haiwan) Bridge

Movement joints
The different structural forms of the cable-stayed sections and the beam sections of the approach roads to the Jiaozhou Bay Bridge are clearly visible. They are separated by movement joints to allow for expansion and contraction between the structural elements.

The navigations channels in the 42.5 km (26⅓ mile) long bridge system (2011) in Jiaozhou Bay on China's east coast are created by three separate structures – one of which is suspended and two of which are cable-stayed. The suspended section comprises a single pylon supporting the twin road deck from its centre, creating two spans. The shortest cable-stayed section comprises two adjacent 'H'-shaped pylons each supporting one section of the twin road deck and creating two equal spans. The longest cable-stayed section comprises two sets of adjacent 'H'-shaped pylons creating one long main span and two shorter outer spans, which are also supported by concrete piers.

Paired pylons (right)

There are four vertical pylons across the width of the twin decks and each half span of the bridge is supported by a total of 24 cables. Distributing the cables between four separate masts allows them to be arranged in a harp design, in which they are parallel rather than fanning out from a single narrower point. It also increases the deck spans between suspension points. Each pair of vertical masts is stiffened beneath the road deck by a horizontal beam supporting the deck, as well as a shared foundation at their base.

Twin deck (left)

The arrangement of the road into a twin deck allows the loads to be divided into two, effectively creating two individual bridges. In the cable-stayed sections, this reduces the size of the pylons and the number and length of the cables. It also spreads the load more effectively into the foundations.

Bridge system (right)

The cable-stayed sections of the Jiaozhou Bay Bridge provide navigation channels in a vast bridge system that comprises the longest bridge over water in the world, including a major intersection in the middle of the bay where three separate branches of the bridge converge.

Glossary

ABUTMENT the foundation structure at each end of the bridge that supports the loads.

A-FRAME TRUSS a truss whose structure comprises two opposing diagonal members joined at their apex and by a horizontal member at their middles, resembling an 'A'.

ANCHORAGE the point at which the bridge's supporting elements, such as suspension cables, are anchored to the ground.

ANCHOR ARM the section of a cantilever bridge that is fixed to the anchorage.

APEX the uppermost part of a bridge or part of its structure.

APPROACH RAMP a section of the deck that provides the link between the road and the main bridge span.

APPROACH ROAD the section of deck leading up to the bridge structure.

BACKSTAY CABLE a major cable on a cable-stayed bridge that is connected to the ground behind the mast or tower.

BALUSTRADE a railing along the side of the bridge, usually comprising a series of upright supports that are topped by a horizontal rail.

BASCULE a counterbalancing mechanism exploited in movable bridges in which the deck rotates about a pivot in the manner of a see-saw.

BENDING MOMENT the sum of forces applied to a beam that cause it to bend. Bending moments create opposing tensile and compressive stresses over the cross section.

BOLT HOUSING the casing surrounding an aperture through which a bolt is designed to sit.

BOWSTRING ARCH TRUSS also known as a tied arch, in which the lateral thrust forces in the line of the arch are taken by a lower chord, often doubling as the bridge deck. Unlike conventional arches, which apply vertical and horizontal thrust to their supports, tied arches apply only vertical thrust.

BOX GIRDER a hollow beam.

BRACE a supporting member that provides additional strength or rigidity in a structure.

BULKHEAD a partition in a structure that can also strengthen it.

CABLE STAY one of a number of independent high-tensile wires extending from a mast or tower to support the bridge deck. Collectively, the cables are arranged either by radiating from the top of the mast, in a fan-like pattern, or in parallel.

CAISSON a pressurised watertight and airtight chamber used in the construction of bridge foundations situated under water.

CANTILEVER a structure projecting in one direction from an upright support.

CANTILEVER ARM the portion of a cantilever bridge that forms part of the main span.

CATENARY ARC/CURVE the natural curve created by a suspended cable supported at both ends.

CHORD the top and bottom longitudinal members in a truss.

CLAPPER BRIDGE a primitive method of bridge construction using only flat stone slabs supported on stone piers.

COFFERDAM a watertight enclosure placed onto the riverbed and drained to permit the construction or repair of a structure. *See also* **CAISSON**

COMPRESSIVE STRENGTH a material's ability to resist compressive forces.

CORBEL a structural element that extends beyond a wall's surface to support members above.

COR-TEN STEEL a patented type of weathering steel that does not require painting and forms a stable rust-like appearance.

CREEP the gradual movement of a material or structure caused by its own mass, applied long-term load or in response to external factors.

CROSS BEAM a transverse beam that connects two principal sides of a structure.

CROSS-BRACED a form of transverse bracing in a structure to increase strength and rigidity.

CROWN the uppermost part of a structure, such as an arch.

COUNTERWEIGHT a weight that acts as a counterbalance.

DECK the main surface of a bridge over which traffic passes.

DOVETAIL JOINT a type of joint comprising an interlocking mortise and tenon.

DYNAMIC FORCE a force that induces acceleration or vibration in a structure.

Glossary

EXPANSION JOINT the space between two parts of a structure to allow for expansion and contraction.

EXTRADOS/EXTRADOSED the exterior curve of an arch.

EXTRUSION the means by which a component (such as a steel wire) is created with a consistent and continuous section.

EYEBAR a structural member comprising a solid bar with holes at either end, forming part of a chain.

FALSEWORK a temporary structure, such as scaffolding, erected in the course of constructing a permanent structure.

FERROCONCRETE a form of reinforced concrete wherein iron or steel bars are cast in concrete to increase the material's tensile strength.

FIN PLATE a sheet of metal welded to a structure to increase strength and rigidity.

FLANGE a section of a structural member that is widened for additional strength.

FORMWORK a mould, usually made of wooden boards, into which concrete is poured.

FRICTION PILE a pile used in foundations whose effectiveness depends on friction with the surrounding soil rather than end bearing.

GIRDER A large beam. Girders are commonly associated with iron or steel, but could be fabricated in any material, such as concrete.

GONDOLA the movable compartment or cabin suspended from cables on a transporter bridge.

HANGER *see* **SUSPENDER CABLE**

HAUNCHED BOX GIRDER a box girder with an underside that arcs from the support to the mid-span. The deepening of the girder over the supports reflects the form of construction and its structural performance.

HELIX a cylindrical spiral or corkscrew structure.

H-FRAME TOWER a structural form characterised by two vertical members joined together at mid-height a horizontal beam, forming an 'H' shape.

HOGGING MOMENT a bending moment that generates tensile forces in the top of a beam, causing the bottom surface to be concave. Opposite to sagging.

HYDRAULIC a mechanism that is operated by the movement of water under pressure.

JOINT *see* **EXPANSION JOINT**

K-TRUSS a truss with two diagonal members between each pair of vertical supports. The diagonal elements, like those on the letter 'K', begin at the mid-point of one vertical and finish at the top and bottom ends of the adjacent vertical.

KEYSTONE the central stone at the apex of an arch.

LATERAL STABILITY providing structural stability through the bridge's depth rather than along its length. *See also* **LONGITUDINAL**

LEAF the portion of a bascule bridge deck that is raised and lowered.

LENTICULAR TRUSS a type of truss comprising two curved chords. The overall outline resembles a lens, from which it derives its name.

LOAD the weight imposed on a bridge. Dead loads are those constant and unvarying forces the structure has to withstand, such as the bridge's inherent weight. Live loads are the unpredictable and varying forces caused by environmental factors or the changing weight of traffic over the bridge.

LONGITUDINAL of or pertaining to the length of the bridge.

MAST a principal structural member projecting upwards, often vertically, from the bridge.

MEMBER a part or element of a structure.

MORTISE-AND-TENON JOINT a simple interlocking joint originating in carpentry where two pieces of wood are joined securely by inserting part of the 'male' component (tenon) into a corresponding hole in the 'female' component (mortise).

PARAPET a low wall that forms a barrier along the outer edge of the bridge.

PERPENDICULAR at right angles to a line or plane.

PIER vertical elements of a bridge that support horizontal elements, such as a beam or truss.

Glossary

PILE a long vertical section of wood or concrete (usually cylindrical in section) embedded in the ground that forms parts of a structure's foundations.

PIN a bolt or similar component that joins two members together by being inserted through apertures in both and fixed securely.

POST-TENSIONED CONCRETE a form of reinforced concrete in which spanning capabilities are enhanced by inducing compression forces in the parts of the structure likely to develop tensile forces. The method involves tensioning steel cables (tendons) within casing sleeves after the concrete has set.

PRATT TRUSS a type of truss in which the diagonal members on each side of the truss are orientated in the same direction and meet at the middle. In a simply supported configuration these diagonals work in tension.

PRECAST a concrete component that is manufactured off-site.

PYLON a tall structural member that acts as a principal supporting element in a bridge.

RADIAL spreading out from a common centre, such as the spokes of a wheel.

REINFORCED CONCRETE concrete that contains reinforcement bars or fibres to enhance its tensile capacity.

RIB a small member used to stiffen or strengthen part of a structure.

SADDLE the mechanism at the top of suspension-bridge towers over which suspension cables are laid. Their purpose is to allow for movement in the cables in response to climatic conditions and functional requirements.

SAGGING MOMENT a bending moment that generates tensile forces in the bottom of a beam, causing the top surface to be concave. Opposite to hogging.

SEGMENTAL ARCH a shallow type of arch that is formed by an arc of less than a semicircle.

SEISMIC of or pertaining to earthquakes or the movement of the earth.

SHEAR FORCE a force acting perpendicularly to the longitudinal axis of a beam.

SICKLE ARCH an arch that resembles the shape of sickle, often being thicker at its mid-section than at its ends.

SLIP-FORMING a construction technique in which concrete is poured into a continuously moving mould, or form. The technique is commonly used in the construction of tall structures with a consistent section, such as the core of a skyscraper or the pylons of cable-stayed or suspension bridges.

SPAN the distance between two supports or piers of a bridge.

SPANDREL the area, often triangular in shape, between the horizontal deck, the curve of the arch (extrados) and the vertical abutment.

SUSPENDER CABLE the cables, often referred to as hangers, that attach the road deck to the main suspension cables on a suspension bridge.

TENSILE of or pertaining to a pulling force i.e. tension.

TENSION *see* **TENSILE**

THROUGH TRUSS a truss-type bridge in which the deck passes through the truss, usually on the lower chord.

THRUST a pushing force acting on or exerted by a structure.

TIED ARCHES *see* **BOWSTRING TRUSS**

TORSION a twisting force

TRANSVERSE BRACING strengthening that occurs across a structure rather than along it.

TRUSS a conventional truss is a structure that derives its strength from the triangle. Each member is subject to compression or tension (or both, though not at the same time).

VAULTING a type of roofed structure formed by a series of arches or vaults.

VIADUCT a type of bridge over land formed by a series of small (usually arched) spans.

VIERENDEEL TRUSS a type of truss formed of rectangular (typically) rather than triangular sections and

therefore relying on the strength of its component members in bending as well as pure tension and compression.

VOUSSOIR one of the wedge-shaped pieces that makes up an arch or vault.

WARREN TRUSS a type of truss made up of only horizontal and diagonal members. Conventional Warren-type trusses have no vertical members, so the alternating diagonal members form a series of equilateral triangles between the horizontal members. Warren trusses with vertical members inserted are called sub-divided Warren trusses.

Resources

Books

*Bridge Engineering:
A Global Perspective*
LEONARDO FERNÀNDEZ TROYANO
(Thomas Telford, 2003)

Bridges: Aesthetics and Design
FRITZ LEONHARDT
(Deutsche-Verlags-Anstalt
GmbH, 1983)

*Bridges: An Easy-read Modern
Wonders Book*
CASS R. SÁNDAK
(F. Watts, 1983)

*Bridges: The Science and Art
of the World's Most Inspiring
Structures* DAVID BLOCKLEY
(Oxford University Press,
2010)

*Bridges: The Spans of North
America*
DAVID PLOWDEN
(W. W. Norton & Company,
2001)

*Bridges: Three Thousand Years
of Defying Nature*
DAVID J. BROWN
(Firefly Books, 2005)

*Bridges of the World: Their
Design and Construction*
CHARLES S. WHITNEY
(Dover Publications Inc., 2003)

Bridges of the World
TIM LOCKE
(Automobile Association, 2008)

*Bridges That Changed
the World*
BERNHARD GRAF
(Prestel Publishing Ltd, 2005)

*Brunel: The Man who
Built the World*
STEVEN BRINDLE,
DAN CRUICKSHANK
(Phoenix, 2006)

*Creation of Bridges: From
Vision to Reality – The
Ultimate Challenge Of
Architecture, Design, and
Distance*
DAVID BENNETT
(Diane Pub Co, 1999)

*Dan Cruickshank's Bridges:
Heroic Designs that
Changed the World*
DAN CRUICKSHANK
(Collins, 2010)

*Handbook of International
Bridge Engineering and Design*
WAI-FAH CHEN, LIAN DUAN
(CRC Press, 2011)

*History of the Modern
Suspension Bridge: Solving
the Dilemma Between Stiffness
and Economy*
TADAKI KAWADA
(American Society of Civil
Engineers, 2010)

Integral Bridges
GEORGE L. ENGLAND,
NEIL C. M. TSANG, DAVID I. BUSH
(Thomas Telford, 2000)

*Landmarks on the Iron Road:
Two Centuries of North
American Railroad Engineering*
WILLIAM D. MIDDLETON
(Indiana University Press,
1999)

*Masterpieces: Bridge
Architecture & Design*
CHRIS VAN UFFELEN
(Braun Publishing AG, 2009)

*Structures: or Why Things
Don't Fall Down*
J. E. GORDON
(DaCapo Press, 2003)

*Superstructures: The World's
Greatest Modern Structures*
N. PARKYN
(Merrell Publishers Ltd, 2004)

Victorian Engineering
L. T. C. ROLT
(The History Press Ltd, 2007)

*What is a Bridge?: The Making
of Calatrava's Bridge in Seville*
SPIRO N. POLLALIS
(MIT Press, 2002)

Web sites

Bridge Hunter
www.bridgehunter.com
Web site database of historic or notable bridges in the United States, past and present.

Bridge Pix
www.bridgepix.com
Web site that includes an ever-expanding database with more than 13,000 bridge photographs.

Bridge Pros
www.bridgepros.com
Web site dedicated to the engineering, history and construction of bridges.

Bridges
www.brantacan.co.uk/bridges.htm
Basic web site that examines the simple and complex structures of bridges providing accurate and informative details. Text-based with illustrations.

Engineering Timelines
www.engineering-timelines.com/timelines.asp
Web site dedicated to celebrating the lives and works of the engineers who have shaped the British Isles.

Highest Bridges
highestbridges.com
Web site detailing in-depth knowledge of 500 of the world's highest bridges. Includes plans, photographs and illustrations.

Historic Bridges
www.historicbridges.org
Web site of photo-documented information on all types of historic bridges in the United States, parts of Canada and the UK.

Nicholas Janburg's Structurae
en.structurae.de
Web site offering information on works of structural engineering, architecture and construction through history and from around the world.

Swiss Timber Bridges
www.swiss-timber-bridges.ch
Web site dedicated to detailing all the wooden bridges of Switzerland, the site currently contains 1389 bridges documented by 19,683 images and documents.

U.S. Department of Transportation: Bridge Technology
www.fhwa.dot.gov/bridge
Government web site containing updated details on bridge technology and bridge studies.

Index

Index

255

AUTHOR ACKNOWLEDGEMENTS

Edward Denison
I would like to thank all the staff at the British Library for their dedication to research and also Ian for his expert advice and direction throughout this project, without which this publication would not have been possible.

Ian Stewart
I would like to thank my wife Louise, for all of her love, support and encouragement as always. I would also like to thank Edward for the enjoyable and creative experience of writing together.

We are indebted to our editor, Caroline Earle, for her patience and professionalism and to everyone at Ivy Press for their invaluable contributions, including Kate Shanahan, Jamie Pumfrey, Adam Hook, and Jane and Chris Lanaway.